D0209128

THE LAUNCH PAD

THE LAUNCH PAD

Inside Y Combinator, Silicon Valley's Most Exclusive School for Startups

RANDALL STROSS

PORTFOLIO / PENGUIN

PORTFOLIO / PENGUIN
Published by the Penguin Group
Penguin Group (USA) Inc., 375 Hudson Street, New York, New York 10014, U.S.A.
Penguin Group (Canada), 90 Eglinton Avenue East, Suite 700, Toronto, Ontario, Canada M4P 2Y3
(a division of Pearson Penguin Canada Inc.)
Penguin Books Ltd, 80 Strand, London WC2R 0RL, England
Penguin Ireland, 25 St. Stephen's Green, Dublin 2, Ireland (a division of Penguin Books Ltd)
Penguin Books Australia Ltd, 250 Camberwell Road, Camberwell, Victoria 3124, Australia
(a division of Pearson Australia Group Pty Ltd)
Penguin Books India Pvt Ltd, 11 Community Centre, Panchsheel Park,
New Delhi—110 017, India
Penguin Group (NZ), 67 Apollo Drive, Rosedale, Auckland 0632, New Zealand
(a division of Pearson New Zealand Ltd)
Penguin Books (South Africa) (Pty) Ltd, 24 Sturdee Avenue, Rosebank,
Johannesburg 2196, South Africa

Penguin Books Ltd, Registered Offices:
80 Strand, London WC2R 0RL, England

First published in 2012 by Portfolio / Penguin,
a member of Penguin Group (USA) Inc.

10 9 8 7 6 5 4 3 2 1

LIBRARY OF CONGRESS CATALOGING-IN-PUBLICATION DATA

Stross, Randall E.
The launch pad : inside Y Combinator, Silicon Valley's most exclusive school for startups / Randall Stross.
vp. cm.
Includes bibliographical references and index.
ISBN 978-1-59184-529-4
1. Y Combinator. 2. New business enterprises—Finance. 3. Venture capital.
4. Entrepreneurship. I. Title.
HG4027.6.S78 2012
658.15'224—dc23 2012018038

Printed in the United States of America
Set in Granjon
Designed by Jamie Putorti

While the author has made every effort to provide accurate telephone numbers and Interner addresses at the time of publication, neither the publisher nor the author assumes any responsibility for errors, or for changes that occur after publication. Further, publisher does not have any control over and does not assume any responsibility for author or third-party Web sites or their content.

For Rebecca, Martin, Jacob, and Alex

CONTENTS

INTRODUCTION

San Francisco Gray Line is the largest sightseeing tour company in Northern California. It offers tours of San Francisco, of Muir Woods and Sausalito or the wine country north of the city, but it no longer offers a tour of Silicon Valley, immediately south. From a bus seat, there just isn't much to be seen.[1]

Silicon Valley's past is more accessible than its present. There's the Computer History Museum, and Intel has a museum of its own. And there are the garages, of course, beginning with Hewlett and Packard's, then Steve Jobs's parents', and then the rented garage that served as Google's first off-campus office space. But these are ghostly places, the uninteresting physical vestiges of startups that have long since departed.

To glimpse what comes next in Silicon Valley, you need to see the most promising startups, not museums and historic garages. There are literally thousands of startups, dispersed along the sixty-mile corridor that extends between San Francisco and San Jose, but they all operate under secrecy until they are ready to launch their first product. That's why there can never be a Gray Line tour of Silicon Valley's future.

It's a shame, because this place is creating everyone's future. *Software is eating the world*—the venture capitalist Marc Andreessen has come up with a rather catchy way of describing the disruption, under way or coming soon, to industries seemingly distant from the tech world. Software-based startups will do much of the disrupting. They take advantage of cloud-based Internet services that make computing power a utility, easily tapped, and whose cost has dropped a hundredfold in the ten years since first introduced. No place in the world has anything approaching the concentration of software startups in Silicon Valley, and this is where, Andreessen says, he expects the majority of future disruptors will appear.[2]

Were it accessible, the single best place to see software startups in the Valley would be the space in Mountain View used by a fund that invests in more software startups than anyone else—dozens at a time. Each company receives a tiny dollop of capital, $11,000 to $20,000, for which it gives up 7 percent of its equity.

Paul Graham, a former entrepreneur, is the fund's chief impresario, its most animated teacher, its chief programmer, and its most prolific essayist, having offered advice to startup founders for more years than the fund, established in 2005, has been in existence. It is he who is most responsible for devising the distinctive features of its operations—such as the absence of hierarchy and titles among partners—and who gave the fund the curious name, borrowed from computer science, of Y Combinator. It is a mischievously obscure choice. It refers to a particular kind of function, for receiving data, performing a calculation, and sending a result back, that most programmers have never used or even understand.[3]

Graham holds a PhD in computer science, but he has no interest in academe. His affinity is for hackers, people who are not merely skilled at coding but have a passion for it. Hackers are curious to know how things work and fix them when they don't. Hackers desire the company of fellow hackers.[4]

Graham is a self-described hacker, and when he launched his own startup about fifteen years ago, the other cofounder was also a hacker. Graham personally wrote the code for Hacker News, an area on the Y Combinator Web site that aggregates links to news stories from around the Web suggested by users who are most interested in programming and software startups and who comment on the stories.

Hackers are, by nature, unruly, says Graham. This sometimes leads to their poking around inside technology where they are not supposed to go. Hacking predates computing, he points out, citing the way physicist Richard Feynman amused himself when working on the Manhattan Project by breaking into safes containing classified documents.[5]

One meaning of the verb form "to hack" refers to finding a clever way to beat a system. Applicants to Y Combinator are asked, "Please tell us about the time you most successfully hacked some (non-computer) system to your advantage." When Y Combinator makes its investment decisions, the teams that Graham most loves to fund consist entirely of hackers.

The YC investment comes with one condition: the team must relocate to the Valley for three months if it is not already there. Those three months—January through March for the winter batch and June through August for the summer batch—are spent working on the product, consulting with Graham and the other YC partners, and coming to weekly dinners with guest speakers. The session—or "batch," which can refer to the session as well as to the group—culminates in Demo Day, at which the teams have the opportunity to make a pitch to several hundred investors.

Individual angel investors have long made seed investments in software startups. But YC pioneered its own twist: mass-producing startups by investing not one at a time but in semiannual batches. Its first batch, in the summer of 2005, had eight recipients. Its twelfth batch, in winter 2011, had forty-four.

These core elements of the YC model introduced in 2005—the batch investing, the three-month residency, the culminating Demo Day—have been copied by dozens of seed funds for software startups that have sprung up around the world. YC remains preeminent, however. It makes more investments than any other seed fund with a residency component; its alumni network, always an important resource to startups that receive investments, is like none other; and investors show the highest regard for YC graduates, informally voting Y Combinator's companies, collectively, as Most Likely to Succeed. YC's principal supplier of capital is Sequoia Capital, the venture capital firm whose roster of successful early investments include Apple, Yahoo, and Google.

A few of Y Combinator's startups have done extremely well. At the top of the list is Dropbox, a file storage service that was born at YC in the summer of 2007 and is now used by tens of millions of people. By installing

Dropbox's free software, a computer user's documents, photos, videos, or any other file that is saved in the Dropbox folder on one PC is automatically synchronized by Dropbox with its Web site and with the user's other PCs, smartphone, and tablet. This makes files always available, no matter where you happen to be. Users get two gigabytes of storage free and pay for additional capacity.

Airbnb, from YC's winter 2009 batch, is another notable. It offers an online marketplace for home owners or apartment dwellers who wish to rent out a spare room (or a couch) to travelers. In early 2011, it celebrated the millionth night booked through its site.[6] Heroku, from winter 2008, is YC's largest success that remains unknown to the general public. It handles the software that runs in the cloud and its customers are exclusively developers.

Y Combinator's reputation for picking the best prospects received validation in early 2011 from two other Silicon Valley funds that joined together to create a fund just for YC startups. They offered a $150,000 investment to every one of the forty-four members of YC's winter 2011 batch. The venture capital world had never seen blanket approval of so many companies in one fell swoop. The YC startups did not have to pass through any additional review process; the fund's investors did not ask to see individual plans or receive any information at all about the recipients. The startups had already passed through YC's own vetting process and had been selected by YC partners as the best among the thousands of applicants. That was all the outside investors needed to know.

Initially, YC was the only seed fund whose portfolio companies automatically received a six-figure investment from outside investors. By the end of 2011, some investors in other places had stepped forward to make similar, if smaller, offers to participants in other programs that funded batches of startups. Marc Andreessen's firm, Andreessen Horowitz, liked the blanket-investment approach so much it decided to invest in every startup in an entire portfolio—and chose to work with YC.

A tour bus disgorging gawking tourists would not be welcomed at YC, but perhaps an author would be permitted to take up residency and observe quietly. More than ten years earlier, I had written *eBoys: The First*

Inside Account of Venture Capitalists at Work, which was somewhat similar to the book I had in mind here. For *eBoys*, I'd observed a then young venture capital firm, Benchmark Capital, and the startups it was investing in. I was resident at Benchmark two years, which happened to coincide with the dot-com boom. The resulting book appeared in March 2000, just as NASDAQ crashed and the boom suddenly approached an unhappy end.

A book about Y Combinator involves far more companies, working at a much faster tempo. In the 1990s, a startup might take a year or two before its founders could turn an idea into a working prototype. By contrast, the YC startups can have something working in weeks, or, in some cases, just a few days. They are encouraged to release something as quickly as possible, discover what needs to be fixed, and try again.

Ten years ago, in order for a typical software startup to be able to attract venture capital, the founders had to have extensive professional experience. They needed millions of dollars to buy expensive servers and database software and to hire employees. Today, YC backs founders who have nothing other than drive and a talent for coding. Still in college and want to start a startup, right away? Just do it. (Startup culture's argot—"start a startup"—ignores redundancy. It also favors fewer syllables over more; one hears a lot about "founders" but never anything about "entrepreneurs.")

The drop in the cost of starting a software business makes it possible for founders to start a startup without capital from YC. But pitfalls await those who try a startup on their own without prior startup experience, or with cofounders who are also innocents. That is one reason why YC has received two thousand applications for each of its most recent batches. Founders seek guidance that will help them avoid the usual mistakes and improve their odds of succeeding. The more one knows about startup life, the more terrifying those pitfalls seem. When YC alumni abandon their YC-backed startup to start over with a new idea, some return to YC to go through the program again.

Even YC-backed startups face long odds, and the founders know it. They often speak of the need to be somewhat "crazy"—a socially sanctioned kind of crazy—in order to ignore those ridiculously long odds and forge ahead.

In thinking about startups, I could see that YC and the founders it backs sit in the center of the Valley's innovation ecosystem. That ecosystem was the best place to observe startups working, out of public sight, on the largest business story of our times, *Software is eating the world.*

The hallmarks of the earlier dot-com boom that preceded the bust—exuberant multimillion-dollar funding of silly ideas and momentarily successful IPOs of money-losing companies—were nowhere to be seen in YC's vicinity. The sums that YC was giving out remained pittances, comparable to what startups in the boom days of the dot-com era spent on Aeron chairs.

None of the middle class's retirement savings was put at risk here. This was an experiment in finance that was being conducted by a small number of professional investors—YC's partners, affiliated angel and venture capital funds, and the wealthy individual investors who would invest after Demo Day.

I wanted to write a book about YC, but only if done the right way, with the twin conditions that business subjects are loath to grant as a pair: providing deep access yet allowing the author to retain unrestricted editorial control. In March 2011, I wrote the YC partners, seeking permission to follow a batch of startups during the upcoming summer. I proposed beginning with the selection process and then continuing with everything that happened at YC during the summer, following as many of the individual startups as I could. My ambition was to give readers a sense of being that unobtrusive observer, the fly on the wall, able to experience what it was like to be there in that place, at that time, amid dozens of startups and half a dozen YC advisers.

Chronicling this would be possible only if I were given permission to work behind the cloak of secrecy that covers YC and its startups. Everything that takes place at YC prior to Demo Day and at each company's individual space is shrouded—no Gray Line tours. Even on Demo Day, a majority of the companies would not yet be ready to have the wider world know of their existence, and the investor attendees would be asked to respect the wish for continued stealth indicated by "off-the-record" notations next to the company's name on the printed program guide. I was asking not just for a day pass but a full-access season pass to this closed world. I pledged that I would use the research only for the book and not use anything in the meantime in my writing for the *New York Times* or other publications.

Happily, the YC partners gave permission for me to do exactly what I proposed. In late April, when the interviews of finalists began, I got started. At the time, I did not know that the summer batch would turn out to be the largest to date, growing to sixty-four companies from the forty-four in the previous one. For me, this was a boon, producing that many more mini-stories—a pageant of creativity and innovation and striving in the Internet age, all found conveniently in one place.

The book originally was going to end at Demo Day, but I have extended the time frame far enough to see one batch's founders greet the founders in the next one. All beginnings are full of possibility, and beginnings are now mass-produced in Silicon Valley. Here is how.

1

YOUNGER

Kalvin Wang and Randy Pang are hackers, the ones who write software. Jason Shen isn't, but he's a self-described "startup sales guy" and has a general-purpose ability to persuade. His personal blog is "The Art of Ass-Kicking."[1] The three are about twenty-four years old, recent graduates of Stanford and Berkeley, roommates living in San Francisco, and fast friends.

They would like to start their own startup together. What the particular business will be is not firmly settled—it changes day to day. In the meantime, they have been selected as finalists who are being considered by Y Combinator for the summer 2011 batch of investments. They are one of about 170 teams—it seems a bit early to refer to them as "companies"—that have been invited to Mountain View for a brief interview. Kalvin and company possess nothing more than an idea for a startup, and a transitory one at that. The "seed" in seed-stage investing doesn't get much smaller than this.

At this early point, the teams do not need to have a name for their proposed company. Most do not. Informally, YC partners will refer to the team by pluralizing one founder's first name or whatever was used as a user name when setting up an online account at the Hacker News Web site, which also receives applications for YC. Kalvin Wang is the principal contact person on the application, so the company is referred to as "the Kalvins."

It's late April 2011 and the Kalvins are led into the small conference room at Y Combinator's headquarters in Mountain View, California. Five men and one woman are arrayed on the other side of the table and along the sides. These are the YC partners, who range in age from a few years older than the Kalvins to two decades older. They all have open laptops and are peering at the application that the Kalvins have submitted online.

The Kalvins know their chat at YC will last only ten minutes. Then a beeper will go off, seemingly almost as soon as the session starts, and they will be ushered out to make way for the next group.

The session will be led by Paul Graham. The other three founding YC partners are present—Jessica Livingston, Trevor Blackwell, and Robert Morris—but interview time isn't carefully apportioned in even increments. It's mostly Graham's show.

The Kalvins sit down. Graham continues to stare at his laptop, then greets them, without preamble. "OK. We liked you guys more than the idea."

The "idea"—the product or service that the startup will offer—often mutates between the time the applicants submit their application and the time of their interview. This is the case here. The idea that the Kalvins had submitted a few weeks earlier was encapsulated as "past memories sent to your in-box." In a preinterview via Skype with a YC partner the week before, they had been encouraged to think of something else.

"We pivoted the idea a little bit," says one Kalvin—Jason Shen in this case, but in the eyes of the YC partners, the finalists are a blurry succession of faces without individual names. "We're going to be the Mint.com for photo books. We organize and rank your Facebook content, allowing you to easily create a printed photo book, featuring the best photos of you, your friends."

Another Kalvin adds, "In college, every dorm—dozens of dorms at Stanford—they have a historian, somebody who makes these books. They have to do this. They usually end up half-assing it. Or they don't finish it. This would be an easy way to get started."

"You guys still have connections in college?" Graham asks.

They did; they had graduated two years earlier.

Graham is much more interested in the founders than in the proposed business idea. When he sees a strong team of founders with the qualities that he believes favor success, he will overlook a weak idea.[2]

"I believe this could be Altair BASIC," he tells the Kalvins, who all were born many years after the first microcomputer, the Altair, was introduced in 1975. "Do you know what Altair BASIC was? Microsoft's first product, right?" Printing photo books could be their start, their Altair BASIC, but Graham wants to know: "What do you expand into from this?"

"We basically believe in the power of memory, nostalgia. We could go into lots of physical products—" a Kalvin begins.

Another YC partner jumps in. Sam Altman is an alumnus from Y Combinator's inaugural batch, summer 2005, and holds a day job as the chief executive of Loopt, the company that came out of that experience.[3] But he's also a part-time YC partner and sits in on the finalist interviews when he can.

"I believe you on the memories and nostalgia," says Altman. "But are people still printing books like that?"

"Yeah. Last year, the photo book printing market was over $1 billion. Europe grew by 25 percent last year."

Another Kalvin adds, "It's been steadily increasing since 2005. The numbers have been going up."

The Kalvins are attempting an improbable thing, making a case for a nondigital product: "Having a physical product that you flip through and have on your coffee table and show your friends—it's really valuable! We've actually bought photo books for our friends and family. It sucks because you have to spend hours making them, finding the photos." Every dorm has to prepare one each year, pay a printer twenty dollars a copy, and buy at least a hundred.

Graham returns to his still unanswered question: "Where does this expand?"

A Kalvin suggests offering a book based on your personal calendar and Foursquare check-ins. Or your tweets for the year.

"You're not serious, that people are going to print up tweets from last year?" asks Trevor Blackwell, who is forty-one, about the same age as the other three founding partners. He too has a day job, as the chief executive of Anybots, the robot company that shares its building with YC.

The Kalvins are unfazed by the generational distance that has left the older YC partners unacquainted with the combination of social media and printed books. "Actually, I have a tweet book," says one. "Tweet books have been going around since 2007. There's a business that basically aggregates the last two thousand tweets you have and prints them out into a book."

Jessica Livingston could picture this. "Kind of like a diary of your tweets," she offers.

"Blurb.com is that book publishing company that went from $1 million to like $30 million in two years. They basically print out blogs, turn them into books," says a Kalvin. "But their software is super clunky—super slow."

"Maybe you can replace yearbooks," Graham says, answering his own question. "High school yearbooks. I feel like it's about time. If you could replace high school yearbooks, that could be a lot of money. It's so clearly waiting for someone to come along."

The Kalvins have made a good impression during their allotted time. As soon as they leave and the door closes, Graham declares, "I like them!"

"We're just not enthusiastic about the idea," appends Livingston.

"Well, it's Altair BASIC," repeats Graham. It's a starting point.

If there is a single perfect age for founders to start a startup, the Kalvins are at that age, a little more mature than undergraduate students but not yet encumbered with the mortgages and children that make leaving a conventional, well-paying job at an established, profitable company so difficult. Still, Graham wants to see in the interview that the team works well together, a prerequisite for surviving the stresses that will come in the life of a startup. The Kalvins have acquitted themselves well, no single one dominating, each comfortably yielding to the others. You couldn't tell which two were hackers and which one wasn't. The one configuration that YC partners are most vigilant about uncovering during the interviews is "hackers in a cage," where one nontechnical founder holds controlling power and treats the hackers as subordinates.

YC has invited less than 9 percent of the two thousand teams that applied to come in for a finalist interview. This means YC partners have brief encounters in person with 170 finalist teams over eight days. In between each interview, the partners type into YC's internal database their individual comments, which the others can see. After each interview, the partners also immediately rank the team among those already seen that day. Funding decisions are made by consensus at the end of the day. Usually the top eight or so teams among the day's twenty-two finalists will receive an offer. These receive a phone call from Graham, who, after a day of sitting, stands and walks around as he talks.

"We want to fund you guys," Graham would say as his greeting when he calls to deliver the offer. "There are three of you, right? It would be $17,000 for 7 percent." The amount of equity that Y Combinator asks for is the same in almost all cases, but the dollar amount of the investment varies slightly: each team would receive $11,000, then an additional $3,000 for every founder beyond the first one, up to $20,000 for four founders.

Most founders accept the offer on the spot, but Graham allows those who wish to ponder the matter to take a bit of additional time. The only responses permitted, however, are yes or no; Graham will not negotiate terms.[4] In a 2008 interview, he described YC's offer as a kind of IQ test for the founders. "If we take 6 percent, we have to improve a startup's outcome by 6.4 percent for them to end up net ahead," he said. "That's a ridiculously low bar. So the IQ test is whether they grasp that."[5]

When Graham finishes his last call of the day, the partners disperse and Graham sits down at his laptop to send an e-mail message to each team that was not selected, with a brief explanation. By the time he heads home, every finalist team that came in that day will have heard back about YC.

Sixty-four companies would be chosen from the two thousand applications (the batch ended with sixty-three after one company decided early in the session to withdraw). This represented a great enlargement from modest beginnings. When Y Combinator had begun six years earlier, Graham's invitation to what he initially called the Summer Founders Program had attracted 227 applications—and only 8 were selected. Graham and his partners were then living in Cambridge, Massachusetts.

It was an "experiment," Graham said at the time, "because we're prepared to fund younger founders more than most investors would." Google and Yahoo had shown that graduate students could start successful startups, he said, and he knew that some undergraduates had demonstrated that they were fully as capable as most graduate students. As the minimum age that was generally deemed acceptable to start a startup was pushed downward, he viewed Y Combinator as a means of finding out just how low that minimum age could go.

Graham addressed prospective applicants who were still undergraduates:

> If you start a startup, you'll probably fail. Most startups fail. It's the nature of the business. But it's not necessarily a mistake to try something that has a 90 percent chance of failing, if you can afford the risk. Failing at forty, when you have a family to support, could be serious. But if you fail at twenty-two, so what? If you try to start a startup right out of college and it tanks, you'll end up at twenty-three broke and a lot smarter. Which, if you think about it, is roughly what you hope to get from a graduate program.[6]

In the spring of 2005, Justin Kan was a twenty-one-year-old senior at Yale. He and his near lifelong friend Emmett Shear, also twenty-one, were working on a Web-based calendar site, called Kiko, in their free time. The largest expense to date was the $250 spent to obtain the Kiko.com domain name. When Graham sent out his invitation, about six weeks before graduation, Kan had not heard of Graham but Shear had, having read Graham's online essays about programming languages and startups. They decided to fill out the application,[7] which could not be dashed off—some of the thirty-two questions were ones they had probably not encountered before:

> **Tell us in one or two sentences something about each founder that shows he or she is an "animal," in the sense described in [Paul Graham's essay] "How to Start a Startup."**

Emmett only learned to program when he was 16 and picked up a C++ tutorial. He did nothing but program, sleep, and eat for the next 6 days.

Tell us in one or two sentences something about each founder that shows a high level of ability.
Justin is graduating with distinction in Physics and Philosophy. Emmett graduated from the Transition School at the University of Washington at age 14, which exists to allow students to finish high school in a single year.

What might go wrong? (This is a test of imagination, not confidence.)
Google might crush us like ants by releasing a superior product tomorrow, supported by their vastly superior backend systems.

If you could trade a 100% chance of $1 million for a 10% chance of a larger amount, how large would it have to be? Answer for each founder. (There is no right answer.)
Justin: $20 million, with sums larger than 100K I am increasingly risk averse.[8]

A few days after e-mailing their application in, the Yale students received word that their application had been rejected.

Thanks for applying to the Summer Founders Program. We're reading the applications now and should have the first round of decisions in a couple days.

We found the applications fell into three categories: promising, unpromising, and promising people with an unpromising idea. Your group was in the latter. We're willing to sit down with groups in this category and work out an idea that might make money, but only if they want to, of course. So would you please let us know whether you are:

a) determined to pursue your current idea

b) willing to consider modified versions of your idea

or

c) willing to work on any good idea, even totally unrelated

Please don't take this personally. It's very common for a group of founders to go through one bad idea before hitting the bull's-eye on the second try. We did. Even Bill Gates and Paul Allen did. (Their first company was called Traf-O-Data. It did not make as much money as Microsoft.)

Thanks,

Paul Graham and Trevor Blackwell

Kan and Shear had been working on their online calendar for five months. How could it not be recognized as a great product in the making? They got over the disappointment, though, and realized they were happy to have been deemed "promising." They wrote back that they would consider modifying the Kiko idea. Shortly afterward, they received an e-mail message, this time from Jessica Livingston, giving them a forty-minute time slot the follow Sunday morning in Cambridge. Their technical competence was not in doubt, she wrote, so the time would be devoted entirely to discussing their business idea.

They got the full forty minutes they had been promised and mounted a defense of Kiko. Then they walked around Harvard Square, waiting to hear whether they'd been accepted. Around six p.m., Shear got a call from Graham. "We'd like to fund you. We'll give you $12,000 for 4 percent of your company."

"I'm not exaggerating when I say we were literally jumping for joy," Kan would recall later. "While it might not seem a lot for anyone who has actually started a company before, from our perspective we were college kids from a school with no culture of entrepreneurship who had just received the external validation that our company was worth 300K! Someone actually thought this was a good idea and we should pursue it!" It was a turning point in his life, Kan said, supplying the means to pursue the startup dream.[9]

✦

Among the finalist applicants granted an interview in April 2011, some are considerably younger than Kan and Shear had been when they'd applied as seniors. A two-person team from Ireland—the older member is twenty-three but the younger one is eighteen and still in high school—has been invited for an interview. When the time arrives, Livingston brings in David Dolphin, the older one, but Patrick O'Doherty is not with him—he remains in Dublin. Dolphin sets his laptop on the table, allowing O'Doherty to be heard via a Skype voice call.

Graham does not regard this to be a satisfactory substitute for physical presence. Y Combinator, through its investments, has made online communication services more robust and varied. But in its own operations, face-to-face communication, without electronic mediation, is deemed indispensable. In Graham's view, there is no way of achieving high fidelity without being physically present. The finalist teams have been directed to come to Y Combinator in person, with all team members present and accounted for. The finalists who are streaming in from around the world, including Iceland, the UK, Denmark, Turkey, South Africa, India, China, Hong Kong, and Australia, must bear the inconvenience and most of the cost: YC reimburses travel expenses only up to six hundred dollars per team.

If funded, all of the founders have to live in the Bay Area for the three-month session. This is necessary because when they consult with Graham and the other partners, it will primarily be face-to-face. Or, in Graham's case, side by side, as he paces up and down the street, with founders at his side, talking as they walk.

All of this is well known to the applicants. Yet here is a two-person team that has failed to get both founders to Mountain View for the interview.

Graham permits Dolphin to proceed, fielding questions about the technology that he and O'Doherty have developed—an Android app used to evaluate the performance of individual cell towers. The two young men think the data that the app collects may be useful to consumers who wish to compare the reliability of different wireless carriers' service. The technical expertise embodied in their testing software impressed him, and even more important, impressed Sam Altman, the resident expert in the room on wireless technology. After a few minutes, however, Graham cannot wait any longer to inquire what possible reason accounts for O'Doherty's seeming defiance of YC's requirement that all founders be present. "Why did you not come to the interview?" he asks of O'Doherty through Dolphin's open laptop.

"At the moment I'm actually supposed to be studying for pretty major final exams at my school, and the parents sort of just axed the idea of coming out," answers the disembodied voice of O'Doherty.

"How would they feel about you doing YC? We're a little dubious about funding eighteen-year-olds. We don't want to screw up people's lives."

"They've pretty much realized that I'll do whatever seems the biggest intellectual challenge," says O'Doherty. "It's not that I particularly want to shorten education. But technology is a good place to be in at the moment."

"You might find it's not as intellectually edifying as you think," says Graham. Outwardly, he withholds encouragement. But O'Doherty and Dolphin have stood up well to the questions. After the interview ends and Dolphin leaves, Graham says, "I like the guys. They pass my test for young founders, which is, they seem older than they are."

✦

After this team leaves the meeting room, one of the partners says something that is said after many interviews: "Fund for the pivot." A "pivot" means a fundamental change in a startup's strategic direction, when one core idea is replaced with another one. Startups pivot all the time and the word is now entrenched in the Valley lexicon of buzzwords. But another, newer term has caught the attention of Jessica Livingston. At one point, in a brief lull in between interviews, she calls out to Altman at the other end of the table, "Sam, you know what my biggest, overused, meaningless tech lingo is? 'On-boarding.' "

Graham, sitting between them, looks up. He hasn't heard that one. "On-boarding?"

Altman recognizes it immediately. "That has just popped up recently—"

"It's driving me bananas," Livingston says.

"Yeah, I don't like it either."

"Do you say, 'How do you on-board users?' "

"Yeah, 'How are you going to on-board these customers?' " Altman says. "I don't ever—that's, I think, the first time I've said it out loud."

✦

A group of four eighteen-year-old college freshmen comes in. They'd received the same sort of qualified rejection that Justin Kan and Emmett Shear had received when applying to the first batch. Once seated, a member of the team gives Graham an update: "After we got the message from you saying, 'We like you guys more than your idea,' we spent a few days brainstorming." They show off what they have come up with: software that displays the full text of articles that show up in a person's tweet stream as links. Graham is not interested.

Lowering his voice to a near whisper, he asks the group, "Why now? Why not wait until you're sophomores? I know it feels like, now, you're going to be over the hill by the time you're sophomores. You're not, you know."

"We don't feel like we're learning."

Another one says, "We're so excited by the opportunity that we're not going to go back." They had worked earlier on an app that showed a user's Facebook friends on a Google map. It had proven wildly popular on their campuses—briefly—but no one seemed to use it after trying it once.

Graham asks, "Do you have any ideas that people would pay money for? 'Cause you know the big mistake that very young founders make is they make some sort of social mash-up thing no one's going to pay for, right?" He continues, "It might be a good form of discipline to force yourself to work on something that someone would actually pay you for. Maybe it's a grubby business, the Internet equivalent of a body shop. But at least it's real."

The beeper goes off. "All right, guys, we're going to contact everybody tonight," he says.

After the group clears out, Graham registers his negative assessment. "I feel like they're very young. They do not feel older than their age."

✦

From the beginning, Graham has been partial to young teams in which all members are hackers. (Justin Kan wasn't a computer science major like Emmett Shear, but he had learned how to code on his own and could list a series of hacking feats on his application.) Among the summer 2011 finalists, the CampusCreds are an anomaly: only one of the three founders is a hacker. The nonhackers are Berkeley students who have not graduated. Their Web site offers college students discount deals from merchants close to their campus.

"You were down in Santa Barbara or San Diego or somewhere like that," Graham says as his greeting, reading the group's application on his laptop. It is hard to tell whether he is saying this in an approving or disapproving tone.

"Yeah, we were down in LA for a little bit," says one CampusCred.

Graham sees that they are nervous. "That was a good sign, getting your hands dirty," he says reassuringly.

He is handed two colorful sheets that show that, the previous fall, CampusCred began offering deals at their home campus, Berkeley, and in the spring semester now under way, the company has added campuses at UC Davis, UC Santa Cruz, UCLA, UC San Diego, and USC. Total revenue: $117,000. Total active users: 11,446. Ten percent weekly growth. Traffic since February: 491,391 page views, 141,364 visits.

"All right, that looks good! What's not to like?" The level of enthusiasm that Graham is showing is most atypical, though the founders in the room do not realize this.

Groupon and other deal sites have little to show for efforts to sell to college students, a CampusCred explains.

"What did they do wrong?"

"They just don't know how to connect to students or properly market their product." The CampusCreds have found that offering students shot glasses, with the CampusCred logo imprinted on the bottom, is a popular promotion. The company has also "partnered" with Greek houses that were willing, for a donation, to drape a large CampusCred banner outside.

Graham cannot contain his excitement. "If I were you, I'd essentially write a program, executed by humans, for how to colonize a new college. And you just colonize all of them. There's no limit! Do them *all* next month!"

The founders laugh, thinking Graham could not possibly be serious, but he is.

Sam Altman, whose voice is always serious, asks the CampusCreds, "Is there a reason you haven't done that?"

"We need a rep at most of the campuses," one answers. It is the rep who enlists local businesses and sets up the deals that are sold on the site.

Altman notices something else that is impressive: at the large deal sites, like Groupon, about 85 percent of the customers are women. But CampusCred's customers are almost evenly divided between women and men. "Why is that?" he asks.

"At Groupon, they do a lot of health and spa. We don't."

Asked to estimate the size of the potential market, the CampusCreds hedge, clearly afraid of appearing delusional. One begins tentatively, "We had some initial predictions of what we would be at if we were at one hundred schools in a year—"

"Why don't you do a hundred schools *a month*?" Graham wants to know.

"It takes a couple thousand dollars to launch a school."

"Are you guys profitable right now?"

"Yes."

"If I funded you guys, I would send you off to investors almost immediately. Get some money and get the landgrab."

After the interview and the CampusCreds have left, Graham is ecstatic: "Now we can calibrate what goodness looks like!"

2

OLDER

"If I had to pick the sweet spot for startup founders, based on who we're most excited to see applications from, I'd say it's probably the mid-twenties," Paul Graham wrote in an essay in October 2006. It was based on a talk he had just given to MIT seniors who were considering whether to go to graduate school, get a job, or, Graham hoped, eventually start a startup.

At that point, Y Combinator had funded three batches of startups. Graham had seen that the youngest founders, who were still in college, had a convenient fallback position if the startup failed. He had come to view the twenty-five-year-olds as best suited to startup challenges—for them, there would be no safe retreat back to school. Startup failure would be embarrassingly obvious to family and friends, which ensured that the founders' commitment to the startup would be total.

Wait too long, though, and prospective founders would be encumbered. Candidates in their early thirties most likely had children and mortgages, which put them at a considerable disadvantage compared to the twenty-five-year-old. "The thirty-two-year-old probably is a better programmer but probably also has a much more expensive life," Graham wrote. The twenty-five-year-old had the most advantages, which included "stamina, poverty, rootlessness, colleagues, and ignorance."

"Rootlessness" meant mobility, which was important if founders happened to live anywhere else but Silicon Valley. They needed to be free to move to the Valley. To illustrate rootlessness, Graham mentioned Kiko, the company that Justin Kan and Emmett Shear had founded in Cambridge in the first batch of YC investments the year before. Kan and Shear were unencumbered with romantic attachments, as far as Graham knew, and were able to easily place all that they owned in a car. At the very moment he was writing, they were on the road, headed for the Bay Area. Rootlessness meant that all of one's possessions either could fit in the car or were not worth moving at all.

By "colleagues," Graham meant former college classmates they had come to know well; these were prospective cofounders. And by "ignorance"—Graham acknowledged that he was deliberately provocative in his word choice—he meant an innocent lack of awareness of the trials that came with doing a startup. It was best for founders not to know what was ahead, or they might not even try.[1]

"Ramen profitable" was one of Graham's best-known precepts in the world of startups. Ramen profitability refers to the moment when a startup, in Graham's words, "makes just enough to pay the founders' living expenses."[2] This was attainable only if those expenses were minimal.

Graham personally knew what startup life was like, from a three-year period, 1995 to 1998, a decade before he founded Y Combinator. The experience ended happily for him: he sold his company, Viaweb, to Yahoo, achieving sufficient wealth that he would never need to work again. But unlike many successful entrepreneurs in the Valley, he did not try his luck again and start another startup. One reason was that he knew all too well how demoralizing the startup experience is.[3] Another reason was that wealth had drained his motivation. On two occasions, he had come close to starting another startup but "both times I bailed because I realized that without the spur of poverty, I just wasn't willing to endure the stress of a startup."[4]

✦

In 1995, on the eve of his adventures as a startup founder, Paul Graham possessed the attributes he would list years later when describing the ideal twenty-five-year-old candidate for a YC investment. He was a hacker, single, had no mortgage or assets, and was earning only enough to cover his expenses. He had stamina, too. Later, looking back on his startup experience, he recalled:

> I also kind of regret being a zombie for several years straight. I really had no life during Viaweb. If people are talking about some famous movie and I've never seen it and have no idea what it's about, it's usually a movie that came out between 1995 and 1998, because at that point, I was on Mars. I was not part of the ordinary world of humans. I was sitting glued to a computer all day long, or asleep.[5]

Graham also did not know at the beginning that this would be his life for three years, so he met that other prerequisite: he was blissfully ignorant.

When he started his startup, Graham fit the profile of the twenty-five-year-old founder perfectly, except that he was thirty. He had a bachelor's degree from Cornell and a PhD in computer science from Harvard, but had no interest in following an academic track. Living in New York, he longed to be a painter and paid his bills as a freelancing software consultant, working for the U.S. Department of Energy, DuPont, and Interleaf.[6] Resolving to start a startup, he hoped to be able to permanently solve that irksome problem of having to work.

Graham dragged his best friend, Robert Morris, a graduate student in computer science at Harvard, into joining the venture. Morris was twenty-nine, but a mishap he had suffered seven years earlier when an experiment of his own design went badly awry had delayed his progress toward his PhD. He had started graduate school at Cornell, not Harvard, where he had received his undergraduate degree. In 1988, in his very first months at Cornell, he had written a little bit of code that would replicate itself so he could count the number of computers that were connected to the Internet. A flaw in the program caused havoc: it replicated itself in an unplanned fashion and the congestion brought down a significant portion of the Internet. The "Morris worm" brought the young student unwanted international notoriety, expulsion from Cornell, and federal charges that could have led to twenty-one to twenty-seven months in prison. He eventually was given probation instead and was able to restart his graduate studies at Harvard.[7] (Without that delay, Graham would later say, Morris would have been a junior professor by the time he reached twenty-nine and "wouldn't have had time to work on crazy speculative projects with me.")[8]

Graham's very first idea for a startup was to offer art galleries a software service to create an online store for their art. Looking back on the experience, he could list many reasons why their company, Artix, became a complete failure, beginning with the fact that art gallery owners had no wish whatsoever to sell art online. Writing about the Artix disaster ten years later, Graham said, "If we, who were twenty-nine and thirty at the time, could get excited about such a thoroughly boneheaded idea, we should not be surprised that hackers aged twenty-one or twenty-two are pitching us ideas with little hope of making money."

Rather than making something that the intended customers emphatically did not want, why not instead make something else? Graham and Morris realized that if they wrote software similar to what they had developed for art galleries and added a "shopping cart," they could create online stores for small businesses of all kinds. For the new company—Viaweb—the two founders secured a $10,000 seed investment and legal assistance from a friend of Graham's, Julian Weber, who in exchange received an equity stake of 10 percent.[9] Morris's roommate was away from Cambridge that summer, so Graham came up from New York and moved into Morris's apartment. Between the two, someone was at work almost around the clock. Morris would begin work very early; Graham would get up at noon and work until four in the morning.

When the summer ended, the two had worked on Viaweb for only one month, but Morris was dismayed that the software was not complete. Seeing that they needed help, they decided to recruit another person, Trevor Blackwell, who was a fellow graduate student of Morris's in the computer science program at Harvard.

Viaweb was not ready to go live until early 1996. About the time when the company had only twenty or so customers, the three went off to New York for their first trade show; when they returned, they discovered that their server, which sat in their apartment, had crashed and had taken their service offline for eleven hours. They were both relieved and dismayed that none of their customers had noticed.

By the end of 1996, Viaweb had attracted only seventy customers, but this reflected Graham's dictum of "get big slow." With few customers, he and his two colleagues were able to make improvements to the software much more easily.[10]

From the moment it was conceived, Viaweb was built to be sold to a larger company. Graham knew that a startup like his either found a willing buyer, which made its founders rich, or it did not find a buyer and eventually would fail.

They entered into many talks about being acquired, but the deals fell through for various reasons. The company came close to running out of funds even as it was negotiating what would turn out to be a solid deal: a $50 million offer from Yahoo, three years after the company began. The company had used only $2 million of capital, so its founders and investors did well.[11]

Back when the company was only a couple of months old, during a dinner, Morris had said he was dubious that anything would come of Viaweb. He made a bet with Graham that were he ever to make a million dollars from the company, he would be willing to get an ear pierced. It would be hard to find a person who seems less likely than Robert Morris to have an ear pierced by his own volition, so this was an emphatic statement of his skepticism. When the Yahoo deal closed, Graham took one arm and Blackwell took the other and they marched Morris to the Garage in Harvard Square, where the unthinkable took place. Morris honored his pledge, Graham reported, but did spend considerable time searching for the smallest earring in the shop.

The sale took place in the bubbly years of the late 1990s. Yahoo turned Viaweb into Yahoo Store. A year after the Viaweb deal, Yahoo acquired Broadcast.com for more than $5 billion, leaving Yahoo with little to show for the purchase but placing Broadcast.com's cofounder Mark Cuban in the ranks of the wealthiest people in the country. Many remember that Yahoo was the source of Cuban's wealth, and Graham is aware of this.

"When you tell people you sold a startup to Yahoo in 1998, they get this knowing look, like you sold someone a bag full of air for a hundred million dollars, but Viaweb was a real moneymaking acquisition for them," he said, a tad defensively, in an oral history of his company.[12]

After the Yahoo acquisition, Graham went to work for Yahoo, which seemed like it would be a congenial place. Its founders had come from graduate programs in computer science and were imbued with the same hacker values as Graham. "They were our tribe of people," he would say.

Graham did not report directly to Yahoo's cofounders, however. He was no longer independent, free to make decisions as he saw fit. He was an employee, reporting to bosses. He spent his days fighting the feeling that he wasn't working so much as constantly deciding whether to submit to someone else's dictates or to rebel. He was not happy and his sojourn at Yahoo did not last long.

The three Viaweb hackers went their separate ways. Morris followed the academic track that he had planned all along, ending up on the faculty of MIT. Blackwell liked the startup life and in 2001 started a startup of his own in California: Anybots, a robot manufacturer. Seven years after the sale to Yahoo, Graham got the idea of setting up Y Combinator, which would make investments but would also allow the original trio to work together again. Graham invested $100,000 and would be full-time; Morris and Blackwell each put in $50,000 and would help in making investment decisions twice a year. They would also be joined by a fourth founding partner, Jessica Livingston, who quit her job at an investment bank in Boston to join YC full-time.[13] She and Graham were a romantic pair and later married; with YC, they combined their professional lives, too.

In April 2005, when Graham announced the establishment of Y Combinator and the plans to select the first batch, the Summer Founders, he referred back to his early unhappy experience at Artix, banging his head against the closed doors of art gallery owners. He expressed his hope that the founders that YC would fund would not waste time or money—YC's money—before learning a key lesson: a company should make something customers actually want. "It would be more convenient for all involved," he said, if YC's founders would "skip the Artix phase."[14]

The best way to ensure this would be to attract applicants who had already produced a product customers wanted enough to pay for or, even better, of course, a product already generating profits. YC's offer to invest a few thousand dollars would be insultingly low for a company that had already released a product that was doing well with paying customers, however. A solution to the problem materialized in the form of two investors, Yuri Milner and Ron Conway, who approached Graham, offering to invest in all of YC's startups on terms that no founder would turn down. The offer caught Graham by surprise, but he recovered and readily agreed.

After the winter 2011 batch was already under way, YC announced the largest change in its history. Every startup was offered a $150,000 convertible note, $100,000 of which was funded by Milner's side and $50,000 from Conway's side. It was called the Start Fund.

For seed-stage investors, a convertible note is attractive because of its simplicity—there is no wrangling over the valuation of the company. It is nominally a loan extended to the startup. If the startup succeeds later in attracting venture capital investors, then a valuation of the company would be set and the note would convert into equity. For the investors, the note is a highly risky loan/investment. By the time a startup discovers that other investors are unwilling to invest in it, it will have spent down the funds from the note and will be unable to repay the loan.

The convertible note was separate from YC's own investment; acceptance of the Start Fund offer was optional. But every startup in the winter 2011 batch snatched up the $150,000 bundle with its name on it. In essence, the Start Fund put chips on every single number on the roulette table, figuring that at least one company in the winter batch would do so well in the future that it would cover the investments in all of the others and still bring a good return.

Now that the Start Fund had appeared, the applicants who were considering applying for the summer 2011 batch would be the first group to pencil in $150,000 in the YC equation. It was not certain that the Start Fund would make the same offer again, but, barring a dramatic change in the investment climate, the odds seemed reasonably good that a present like that given to the winter 2011 batch would be given to this next batch, too.

✦

If you and your cofounder are in your thirties and married, with young children and mortgages, you have the complications that Paul Graham and Robert Morris did not have when they started Artix and Viaweb. If you two collectively have five children five years old or younger and your spouses do not work outside the home, it would be all the more difficult to drop your full-time jobs and give the startup your complete attention. And if you lived in Birmingham, Alabama, you're starting at a considerable remove from Silicon Valley.[15] Jason McCay and Ben Wyrosdick were well aware of all this. Their chances of landing even an interview with YC seemed so slim that they wondered whether they should even bother preparing the application, which required posting a one-minute video in addition to answering many written questions. But Wyrosdick was optimistically inclined by temperament and wanted to try. McCay, the pessimist, had to be persuaded to go along with him. He recalled when, as a college student in 2000, he had applied for a position with Sapient, a fast-growing consulting company whose services were in great demand by technology companies. He'd been granted an interview slot in New York and sent a plane ticket to fly up. He was elated—briefly. When he arrived, he discovered that the other candidates were mostly drawn from the Ivies. "Where are you from?" he was asked again and again. When he answered, "Auburn," he got puzzled looks and the same recurring question: "What state is that in?" Some fellow candidates were friendly, but one from Harvard was insufferably condescending. McCay had not expected the outcome: Mr. Harvard did not get an offer—but he, Mr. Auburn, did.

In 2011, McCay and Wyrosdick, who also went to Auburn, knew that their background did not conform to an ideal template for founders the way others would—say, recent Stanford graduates who were in their mid-twenties. But they thought that if they managed to get an interview at YC, that itself could be considered an accomplishment.

The two had established a growing software business, one that offered an online service that software developers clearly wanted. Their company provided an easy way for developers to set up a relatively new database product, called MongoDB, on servers that were in the cloud. MongoDB software came from another company, 10gen, which, rather than selling it, offered it free as open-source software; 10gen sold consulting and training services.

McCay and Wyrosdick offered MongoDB to database developers who did not wish to install MongoDB on their companies' machines but instead preferred to pay for the convenience of a cloud service. It originated in their own experiences as employees and software consultants; they had wanted someone else to set up and handle the administrative headaches that came with keeping the MongoDB databases running. But 10gen had not originally designed MongoDB to be installed in the cloud; developers were expected to install it on their own servers. What McCay and Wyrosdick wanted—MongoDB hosted in the cloud as a service—simply did not exist. So they created it, figuring there would likely be some other developers who would want something like it, too. Perhaps they could build a sideline business out of it and make a little money. They named their company MongoHQ, which left no question what line of software they were in. They worked on MongoHQ in the evenings and the wee hours of the morning while holding down day jobs.

They took heart from "Selling Pickaxes During a Gold Rush," a blog post published a couple of months earlier, in February, by Chris Dixon, a seed investor who was based in New York City but well known and respected in Silicon Valley.[16] During the California gold rush, some of the most successful businesspeople—like Levi Strauss—didn't mine for gold themselves but did well selling supplies to those who did. Today, Dixon argues, entrepreneurs who use the latest technology face a similar choice: they can sell to consumers—what Dixon calls "mining for gold"—or they can sell the software tools that other developers would use to create the consumer product—that is, "selling pickaxes." Dixon mentioned that Y Combinator's most successful "exit" to date was Heroku, the company that sold cloud-related services to other software companies, the digital era's equivalent of selling pickaxes to miners. It was sold to Salesforce.com less than three years after its birth at YC for more than $200 million cash.[17]

The two MongoHQ founders cited the Dixon post in their YC application and declared, "We are selling pickaxes during a gold rush."

In the section where they were asked to explain how the founders had met and how long they had known each other, they gave answers that would be far more likely to come from older founders than younger ones: they had known each other ten years; had worked together for much of that time; and could write, "We have often been referred to by others as work wives, although we argue about who is the wife in the arrangement. :)"

When they applied, they could report that MongoHQ had 5,100 accounts and about $5,500 of recurring monthly revenue. Its user base was growing about 16 percent monthly. They did not realize just how far ahead of most of the applicants they would be, the ones who had nothing but an untested idea for a product based on what the founders imagined was a need. Whether it should be attributed to their longer experience as developers or not, McCay and Wyrosdick had selected an idea that had the highest probability of market success: it fulfilled a professional need that they themselves had.

Heroku hosts the code that runs Web sites (initially, it supported Ruby but later added other programming languages). MongoHQ offers a complementary service, supplying the database needed by the program that runs on Heroku. The MongoHQs have a revenue-sharing deal with Heroku, which offers its customers the option of setting up MongoDB databases that are overseen by MongoHQ.

McCay and Wyrosdick didn't realize how important the relationship with Heroku would be when they applied to YC. Adam Wiggins, one of Heroku's cofounders, told Paul Graham he had worked with the two founders and thought highly of them. So while the two MongoHQs thought of themselves as the two unknowns, they had, in fact, been moved before their interview to the highest-ranking tier of candidates, occupied by the applicants who had been either YC alumni themselves—there were a number of returnees who wanted to go through YC again with a new startup idea—or those whom the most successful alumni had personally recommended.

✦

The MongoHQs are the first interview of this day in late April. Jessica Livingston ushers McCay and Wyrosdick into the interview room, everyone shakes hands, and the two take their seats. As is his wont, Paul Graham begins with no more preliminaries than a brief "OK" and then reads off from their application the number of accounts that MongoHQ has. McCay has an update: the number is now much bigger.

"That's good!" Graham laughs. "When the number's bigger than it was at the application." He lets them know that their application has been bolstered by an important recommendation. "Adam Wiggins really liked you guys."

McCay says they enjoyed working with Wiggins when they integrated MongoHQ with Heroku. Today, about 30 percent of MongoHQ's accounts are on Heroku.

"They have this new vision for the world," Graham says, "where everything's managed services. Databases hasn't been a service before."

Trevor Blackwell jumps in, with some not so friendly sounding questions. "Why is it a good service? Why don't I set up MongoDB myself?"

"The setting-up part is not the hard part," says Wyrosdick. "It's kind of the going forward. I guess it's the same with Heroku. You set up a Web server pretty easily, but it's working out those pains all along the road. Scaling, monitoring, everything that goes along with it. Who's up at three in the morning when those things aren't responsive?"

"Are you?" Graham asks them.

"We are."

"Really? You do that?"

"We get alert phone calls," says McCay.

MongoHQ does not own and maintain its own servers. Its service runs on Amazon's Web servers, its EC2 service. Amazon's developer services had experienced an embarrassing outage earlier that week.

"Did you have a lot of excitement this week with the outage?" asks Blackwell.

"We did," McCay says equably. He pulls out a sheet with a graph. "I wanted to show you this—it's just our growth. The red line is our databases being added, and the blue line is our monthly recurring revenue." The lines head in a pleasingly upward direction.

Graham contains his reaction until the interview is over, the two founders leave, and the door is closed again. "It's rarely that easy!" he says, then corrects himself: "Actually, it's rarely that easy in a good way. Sometimes it's that easy in a bad way."

The pair of thirty-somethings from Birmingham, Alabama, are in.

3

GRAD SCHOOL

Around midmorning on the last day of April, after six days of finalist interviews, selections, offers, and acceptances, Paul Graham addresses about one hundred founders who have been admitted for the summer batch. They fill YC's main hall, which is not large enough to accommodate the entire batch. Two more days of interviews and acceptances will follow, with a repeat of the kickoff meeting. YC has outgrown its current space—walls must be moved to expand the hall before the dinners begin in a month—but in the meantime, only three-fourths of the future class can be assembled in one place.

During the interviews, the Paul Graham seated across the table from the founders had not taken pains to seem welcoming. He did not need to—in the division of duties, Jessica Livingston was the one who ushered each group into the room from the main hall, and she had a thousand-watt smile for everyone. She walked the teams in and announced the primary contact person's user name. Then Graham, who had been relieved of responsibility for pleasantries, got the questioning under way immediately. His manner and the questions themselves often had a sharply skeptical tone. Today, however, the Paul Graham who stands before the large group seems more benign. He is wearing the informal cargo shorts and sandals that hackers wear. If the temperature drops, he'll add a sweatshirt but will keep the shorts and sandals. For the kickoff meeting, he doesn't start off with a joking preamble or formal welcome, but the adversarial tone is gone. He looks down at notes, looks up and makes a point, and then dives back into the notes.

"There are forty-eight startups so far. It'll probably end up being sixty total, roughly," Graham tells the group. "Out of about two thousand applications, so that's 3 percent. Which is about normal."

"You guys are the thirteenth batch. After we finish picking, we're definitely going to be over three hundred startups. We're more than halfway to the name of Dave McClure's thing." This was a reference to a Y Combinator competitor that was close by: 500 Startups, a startup incubator and seed fund in downtown Mountain View that was founded by McClure in 2010.

Graham introduces YC partners and the attorney on retainer with whom the founders are invited to consult anytime they would like. Then Graham offers some advice: "In general, don't hide your disasters. We're not going to take the money back." He says this lightly, as if delivering a joke, but it is reassuring for the founders to hear. They laugh, perhaps with a touch of relief.

The $150,000 that the Start Fund had offered to the last batch? It is offered to this group, too, he says. There are a few tweaks but it remains a very founder-friendly form of financing. "You'd still be crazy not to take it," he says.

The founders should be thinking about Demo Day, right now. It is the chance to "mass-produce reputation," Graham explains. "Everything you do between now and Demo Day can affect how good you seem on Demo Day, right? And everything you do one hour after Demo Day will have *no* effect on how good you seem on Demo Day."

The YC dinners on Tuesdays are not mandatory. Once the founders are living in the Bay Area, nothing is mandatory. "Unlike many investors, we have no power over you. So all we can do is beg you," says Graham, pleading that the founders live in Mountain View, near Y Combinator. The middle of Silicon Valley is boring, he knows. But he says:

> If you live in San Francisco, it will introduce drag on your coming here, and that will cause you to occasionally miss things. We don't

do things here just for our amusement. We do it because it's actually useful. Like we'll have some famous person who suddenly can speak but it turns out only on a Wednesday. And you're going to be thinking, 'I just came down for dinner. I need to do some work. I need to drive down to the Valley?' You wouldn't have to have that decision to make if you lived here. You'd just bicycle over. I know: you guys don't want to live in Mountain View. Do it for three months. After you're done, you're funded, move to San Francisco and be hipsters.

He tells the founders that they are not going to be closely supervised, that they should not assume that Y Combinator partners will hover close by and save them from screwing up. "A lot of you guys—this is your first startup. You've been sort of trained by circumstances, by school and work, that there is going to be someone who will nag you if you do badly. If you're not delivering, your boss will say, 'Hey, you're not delivering,' and eventually he'll fire you if you're not delivering. Here, we don't fire you. The market fires you. If you're sucking, I'm not going to run along behind you, saying, 'You're sucking, you're sucking, c'mon, stop sucking.' "

The group laughs.

"I tried it a little bit, out of frustration. But it doesn't work. When people are doing badly, they continue to do badly. It's kind of sink or swim. We'll help you. We love to help you if you want help. But if you drift off and do nothing, we're not going to come find you and drag you back."

The most successful startups, Graham says, are the ones that completely remove distractions: "They just sleep, eat, exercise, and program." He gives the example of the Zenters—Zenter was a startup that created presentation software, like PowerPoint, but the software was used at the Zenter's Web site rather than downloaded and installed like Microsoft Office. The Zenters were in YC's winter 2007 batch. "They got this apartment together a couple blocks from here," says Graham, "and just got a bunch of Lean Cuisine and put it in the freezer, and they programmed and occasionally played tennis and ate Lean Cuisine. That was what they did. You can bear it for three months, or however many days it is until August 23."

Standing to the side of the room, Jessica Livingston calls out, "They each lost fifteen pounds!"

Graham points to the Zenters as exemplary models not only because they worked diligently but because for a long while they held a YC record for being acquired most quickly. (Google acquired the company just two months after Demo Day and used Zenter's technology for its Google Apps presentation software.)

As Graham moves through the list of points on his sheet of paper, it is easy to miss the significance of one of the points, the one in which he lightly predicts that "more than half of you will fail." He says he would be delighted if the percentage of startups that succeeded were as high as 50 percent. But it's unlikely to happen.

It's not clear what he means by "succeed." When the founders permanently solve their "money problem" as Graham had done with Viaweb? Or simply struggle on?

The modest amounts of capital that YC invests in individual companies make it possible to wonder about such questions. There is nothing in the manner in which YC operates that suggests that Graham and his partners are only looking for one Dropbox and do not care about the others. But just as getting into YC is unlikely—a 3 percent chance, as Graham pointed out at the beginning—achieving notable success, even as a YC company, might be even more unlikely. Perhaps there is no better than a 0.3 percent chance of that. Graham has not released data about the fate of earlier YC investments, so there is no way of knowing just how low the percentage of successes actually was.

"You should start now," he tells the founders. The dinners would start in a month but office hours could begin immediately. "Y Combinator already started—you just didn't realize it. Start doing what you'd be doing."

Before they head off, the new founders are invited to mingle in the hall. "Everybody else doesn't know anybody here, either," he says. "This is like the first day of college. Introduce yourselves to as many people as you can. Figure out who your friends are going to be in the batch."

They stay, turning to neighbors and introducing themselves, their voices bouncing off the concrete floor, creating a din. Graham was right—it *is* like the first day of college. Or, more precisely, like the first day of graduate school. YC will resemble, in many ways, the last portion of a PhD program, after course requirements have been fulfilled and oral exams passed, and students are working on their dissertations. Graduate students work mostly off campus and meet their advisers only occasionally—when the students choose. Occasionally, there are social events with fellow grad students that pull them out of the isolation of the library stacks or lab. How a student chooses to spend his or her time during any given day is wholly up to the individual. That's how YC founders work, too, during their three-month residency.

Journalists are fond of referring to Y Combinator as a "boot camp." But it is nothing of the sort. Founders are not stripped of their individual identities. There is no ritualistic equivalent to buzz cuts or uniforms, no punishing regimen that everyone has to go through, no obstacle courses to build team bonds and respect for a chain of command. There is no formal program or curriculum of any kind. Instead, the only regularly occurring part of the experience is the weekly dinner on Tuesdays, served out of Crock-Pots and rice cookers. After dinner a guest speaker will present an off-the-record talk about his (all the speakers this summer are men) startup experiences and lessons he has learned. Before, during, and after dinner, the founders will talk with one another about their most recent progress and current headaches, and they can consult with Graham and the other YC partners. Graham believes having the founders gather once a week, in person, prods them to work harder because of the power of shame avoidance: they would not want to embarrass themselves by having little progress to report to their peers at the weekly get-together.

Graham and his YC partners insist that the founders be physically present for the consultations. When one founder asks if he could use Skype to have a video chat in place of coming over to YC for a conversation in the flesh, Graham says he'd permit it but then frowns. "It doesn't work as well," he says. "It turns out that startups are so hard, you kind of have to talk about them face-to-face."

To organize those face-to-face conversations, YC has adopted another academic practice: office hours. In the earliest batches, there was no need to set up designated times to talk with founders. Graham himself did the cooking for the Tuesday dinners and founders simply came over and chatted while he worked in the kitchen. When the number of founders in a batch grew too large, Graham handed off cooking duties to professionals and put office hours in place. But he runs his differently than on a campus. Typically, a faculty member's office hours are available for drop-in visits; no appointments are necessary. Graham, who could not abide wasting time, built software for an appointments-only system. Founders who sign up online are automatically assigned to a particular twenty-minute slot; no block of time is left idle. Founders can have many consultations, or few, or none at all. It is entirely up to them.

The similarities between graduate school and YC struck Matt Brezina, who, as a cofounder of Xobni, was a member of the YC batch of summer 2006. Five years later, he composed a blog post: "YC: The New Grad School."[1] Brezina had been a graduate student at the University of Maryland before he left to start Xobni. In his view, Graham ran YC not to make money but to teach. It wasn't a stretch. Anyone who had grad school experience and watched Graham during his office hours with YC founders would think of Graham as a highly engaged teacher, albeit one with an intensity not always found in a faculty member's office.

Brezina claimed Graham had created a new type of "university," with weekly lectures (Tuesday dinners), a practicum (founders launch a product), and readings (Hacker News, the Web site created by Graham that hosts discussion of tech stories suggested by its users and that has a wide following among coders).

The residency requirement of this "university" rankled some in YC's first batches. Responding to the criticism in 2006 when YC funded a dozen companies at a time, Graham said even if he were personally able to travel to twelve scattered locations around the country, it would not be in the best interests of the startup to stay where it was:

> We wouldn't be doing founders a favor by letting them stay in Nebraska. Places that aren't startup hubs are toxic to startups. You

can tell that from indirect evidence. You can tell how hard it must be to start a startup in Houston or Chicago or Miami from the microscopically small number, per capita, that succeed there. I don't know exactly what's suppressing all the startups in these towns—probably a hundred subtle little things—but something must be.[2]

The complaints about YC's requirement that founders relocate to the Bay Area diminished as the years passed, however. Even though the technology for video chat improved, no one could credibly argue that it was equivalent to a conversation in person. This was especially the case if the conversation included Graham, whose métier was speaking one to one or one to two, when he was excitable and demonstrative in ways he was not when speaking to a larger group or facing a webcam.

It would have been more difficult to convince YC applicants of the desirability of moving to where YC was had it stayed where it started, in Cambridge. But after the first YC batch had finished in August 2005, Graham decided it would be a good idea to plant the flag in Silicon Valley. He expected that other seed funds would pop up and copy the Y Combinator model, and he didn't want to leave open an opportunity for someone to come along and say, "We're the Y Combinator of Silicon Valley." He wanted Y Combinator to *be* the Y Combinator of Silicon Valley. He and Livingston planned to move out to the Bay Area for the next batch, in winter 2006, and then alternate, running a summer batch in Cambridge and a winter one in the Valley.

They would need one room large enough to seat everyone in the batch for the weekly dinner, but only for a few months every year. Trevor Blackwell offered to carve out sufficient space in Anybots' building in Mountain View.[3] The space included one small conference room with a door that closed, which was used for the interviews, and an open kitchen, where the dinners were prepared. This adjoined a high-ceilinged room large enough to accommodate one thirty-foot-long table and benches, large enough to seat everyone in the batch.

There was nothing about the facilities or the location that resembled a sylvan college campus. The street happened to be named Pioneer Way and it was located in a light industrial area, a forlorn triangle bordered on two edges by highways. It was about a twenty-minute drive from where the posh offices of venture capitalists on Sand Hill Road were concentrated, near Stanford, on the leafy west side of Menlo Park. The neighborhood of Pioneer Way belonged to a separate galaxy. YC sat among small manufacturers, and auto repair and body shops. The architecture in the neighborhood was strictly no-frills utilitarian—a good setting for lean startups. YC was there every other winter until 2009, when Graham and Livingston decided to make the Valley their permanent home and run the program there for both the winter and summer batches.[4]

Two years after YC's founding, a seed fund named TechStars sprang up in Boulder, Colorado. It presented the first real competition to Y Combinator. Founded by David Cohen, Brad Feld, David Brown, and Jared Polis, TechStars made its first investments in summer 2007. It shared many characteristics of YC and invested a similarly tiny amount of seed money—less than $20,000[5]—in a batch of startups. It took a similar slice of equity, 5 percent originally, later 6 percent. Like YC, TechStars required that founders were to be resident for three months and it organized a Demo Day for investors at the end. Its application included the questions used on YC's application.[6]

TechStars did differ from YC in several respects. TechStars' batches remained small, only ten startups in each, plus or minus one. Instead of having full- and part-time partners, as YC did, who were available to all of the startups in a batch, TechStars arranged to have volunteer "mentors" on hand: investors or experienced entrepreneurs, a set of whom would work with one particular startup. This too set TechStars and YC apart: Graham never used the term "mentoring."

Other seed investors joined together in many places to create similar programs, providing funds to a batch of startups and structured guidance. Strictly speaking, programs that provided work space were "incubators" and those that did not were "accelerators." But the terms were sometimes used loosely. The others all shared one characteristic, however: they were singletons. Chicago had Excelerate Labs; Durham, LaunchBox Digital; San Francisco, KickLabs; Dallas, Tech Wildcatters; Pittsburgh, AlphaLab;

Cincinnati, The Brandery; Austin, Capital Factory; New York, NYC SeedStart; Providence, Betaspring; Salt Lake City, BoomStartup—and this is but a sampling.[7] Only TechStars quickly built out multiple locations. To its summer session in Boulder, it added a winter session in Boston, a spring session in New York City, and a fall session in Seattle. The total number of startups funded by TechStars in calendar year 2011 was still only about one-third the number that YC funded. But spreading its program into four cities brought TechStars more attention in tech blogs than the singletons could. TechStars also helped organize in 2010 the Global Accelerator Network, a federation of accelerators around the world. YC did not join the network, nor did it show any inclination to open a second branch elsewhere.

TechStars was nominally an accelerator, not an incubator, because it did not formally require teams to use the free space it offered. But almost all did so, regarding the working in close quarters with nine other companies in one room as a benefit. "Definitely intense," said a TechStars alumnus about the summer 2011 session in Boulder.[8]

By design, Y Combinator did not provide work space for its companies. Graham had a strongly negative opinion of incubators. He recalled that when he and Robert Morris had started out with Viaweb, they never would have accepted funding from an incubator. "We can find office space, thanks; just give us the money," he wrote later. He believed that an independent cast of mind, which would chafe under the close supervision in incubator-supplied office space, was needed for success as a startup founder.[9] He sought those with the same outlook when selecting founders for YC.

Everyone at YC—founders and partners—spent most of the week dispersed, working independently and coming together, usually just once a week. Graham and his partners did not have room for their own offices at YC. They commandeered a tabletop in the main hall when they needed to use their laptops. Y Combinator fit the definition of accelerator, but Graham was fussily averse to calling YC anything other than plain "seed fund."

TechStars played up its own openness, contrasting it with Y Combinator, which did not disclose information about whom it funded and how they were doing. "At TechStars, we believe in full transparency," said the challenger, listing every company in every batch, how much capital each had raised, and whether the company remains active, has been acquired, or has failed. Y Combinator adopted the position that it could not give a full accounting of each batch because doing so would entail the release of information that at least some of the startups had not yet released. But YC did have some companies that were doing conspicuously well, like Dropbox and Airbnb. And there was Heroku's $200 million-plus exit. With the then one-of-a-kind Start Fund, Y Combinator was seen in spring 2011 as the most desirable seed investor to founders with the greatest ambitions.[10]

One such pair were Michael Dwan and JP Ren, the cofounders of Snapjoy, a startup building a Web site that would store photos and organize them in novel ways. The two were hackers in their mid-twenties who were accepted into TechStars' summer 2011 batch in Boulder. They should have been pleased: the two lived in Boulder, a place they had moved to and that they loved. They had no desire to budge, but they did not want to follow TechStars' curriculum. From what they could see, TechStars would have them spend the first month of the three-month program wooing mentors and the last month pitching investors, leaving only the middle month to work full-time on building the actual product.

Snapjoy was planning a site that could handle many thousands of photos for each user—entire hard drives filled with photos, every single photo the user had ever taken. "Doesn't Picasa already do that?" a TechStars partner had asked them during their interview. Yes, Dwan said, and we plan to kill Picasa. And Flickr. And iPhoto. It seemed to Dwan that the TechStars panel was confused by his and his cofounder's ambition.

The Snapjoys were also irked when they were asked during the interview, "So who's the CEO?" CEO? There were just the two of them. "Why do we need to create this artificial hierarchy?" Dwan said when recalling the exchange.

TechStars ran its application process for its summer session early, and applicants who were accepted had to commit to TechStars before they would hear from Y Combinator. Having been accepted into TechStars and been given forty-eight hours to accept or decline the offer, Dwan and Ren were in a quandary. They had been given a YC interview slot, but that was still two weeks off. So they arranged with YC to fly out immediately to do the interview and were accepted on the spot.

Michael Dwan's wife came out to Silicon Valley with him for the summer. But she came to be a spectator and startup widow, not a founder. Ren faced a summer alone, as his girlfriend decided to stay in Boulder. But the cofounders' worries about whether they had made the right choice fell away within the first few minutes of Paul Graham's remarks at the kick-off meeting. He said to the group: if you're not fully focusing on your product to the exclusion of all else, you're wasting your time.

Dwan turned to his cofounder and smiled. That's why they'd come.

4

MALE

In Paul Graham's view, startups are qualitatively superior to large corporations in just about every way. Leading the list of the startup's superior attributes is the ability of its founders to choose one another and then hire employees considering nothing but merit:

> One advantage startups have over established companies is that there are no discrimination laws about starting businesses. For example, I would be reluctant to start a startup with a woman who had small children, or was likely to have them soon. But you're not allowed to ask prospective employees if they plan to have kids soon. Believe it or not, under current U.S. law, you're not even allowed to discriminate on the basis of intelligence. Whereas when you're starting a company, you can discriminate on any basis you want about who you start it with.[1]

This was March 2005, when Graham addressed the Harvard Computer Society. He spoke as a retired startup founder; he was also speaking when he did not yet have children of his own. (It was later that same evening, after the talk, that he would decide to unretire and establish Y Combinator.)

These words would be read back to him years later by critics of YC. To them, it appeared that Graham was discriminating in the same way that he had described in the essay when it came to selecting YC participants. The evidence seemed clear enough. YC's classes were almost all male.

✦

For Jessica Livingston, the only female among YC's partners, the absence of women among YC's startup founders reflected the dispiriting lack of women founders in tech. She was well aware of the pattern. She had started work on *Founders at Work* before cofounding YC. Among the book's thirty-two profiles of successful software entrepreneurs, only three feature women: Caterina Fake of Flickr, Ann Winblad of Open Systems and Hummer Winblad and Mena Trott of Six Apart.

YC doesn't ask about gender (or race) in its application and does not have an exact statistic for the male/female ratio. But up through its twelfth batch, the winter 2011 class, Livingston estimated that only about 4 percent of the founders had been women. Then came the summer batch. Among its 160 founders, instead of the 6 or 7 women founders that would be expected based on the male/female ratio in previous batches, there were only 2 women. Two out of 160 is just a hair more than 1 percent.

One was Elizabeth Iorns, the chief executive of Science Exchange. Her company was preparing a marketplace for university-based scientists, matching those who lacked the equipment for a particular experiment with those who had the equipment and the willingness to take on contract work. In the world of software startups, founders are customarily divided into two groups: "technical," which means coders, and "nontechnical," which means everyone else. YC heavily favored teams with a majority of, or nothing but, technical founders. In the case of Science Exchange, however, Iorns was an anomaly: she held a PhD in cancer biology, specializing in the biology of breast cancer. She was not a hacker—the other two founders, her husband, Dan Knox, and Ryan Abbott, were—but she was not a marketing person or other business school type, either.

The other woman in the summer batch was Yin Yin Wu, one of the three cofounders of Adpop Media, which offered software to video producers and advertisers that enabled products or brand names to be digitally placed into videos with an uncannily natural appearance, as if the objects had been present when the video was originally shot. Adpop was extending work that had been done at Stanford's Artificial Intelligence Lab. Wu and her two cofounders, Shazad Mohamed and Xuwen Cao, were all hackers.

When they applied to YC, the Adpop founders were taking their last classes as seniors at Stanford. The YC summer session started before they had taken their final exams, but any difficulties this created were overcome by dispensing with unnecessary things—namely, sleep.

Mohamed, who had grown up in Dallas, started programming in middle school. Wu had begun programming later, in high school, in Louisville, Kentucky, and only incidentally, in order to work on her science fair projects in computational chemistry. Ask her what she was working on then and she explains, "Modeling proton transport between crystal channels."

Wu arrived at Stanford thinking that she would major in biology or chemistry and took computer science classes as a side interest. "The CS classes were the ones that I most willingly spent time on," she tells me. "Even when I finished the assignment, it's like"—here, she jokes—" 'Dude, what else can I hack out?' "

She also noticed that computer science could be used in other fields. When she was doing computational chemistry, she was the only person in the research lab who could program. For her, the programming was relatively simple, but her nonprogramming colleagues were in awe.

In Stanford's class of 2011, about ninety graduates out of the sixteen hundred were computer science majors. Of those ninety, about 20 percent were women. This sets a benchmark. Wouldn't YC, with its preference for young hackers, end up with a contingent of women who made up a similar percentage among its hacker-founders? Why was there just one Yin Yin Wu?

It was a question that occurred to the men of the class of summer 2011. At the second dinner, when Yuri Milner, the guest speaker for the evening, was fielding questions from the group, a male founder called out, "Why are there no women here?" This wasn't a question that Milner could answer—he had no role in the selection of companies for YC. Livingston, standing at the side of the hall, called out to the founder, "Why are you asking *him*?" Graham, standing at the rear, quickly piped up and said he would supply an answer. There was no great mystery: the absence of women reflected the absence of women among those who had applied.

Some female founders had come in for interviews as finalists, but they were not hackers. The problem, Graham suggested, was that hacking generally begins around the age of thirteen, when boys show more interest than girls. He also observed that hackers tended to spend time with each other as pairs, and the pairs were same sex. In the six-year history of YC, with more than three hundred companies funded, there had been only one instance in which there had been an all-female team, a pair of cofounders.

✦

Paul Graham has been asked the same question about the absence of women since the earliest days of YC. In October 2005, after the very first batch had gone through for the initial run, he noted in an essay that the shortage of women was manifest not just at YC but also at startups generally. He had recently run across mention that only 1.7 percent of venture capital–backed startups were founded by women. "The percentage of female hackers is small, but not that small. So why the discrepancy?" he asked. "When you realize that successful startups tend to have multiple founders who were already friends, a possible explanation emerges. People's best friends are likely to be of the same sex, and if one group is a minority in some population, pairs of them will be a minority squared." Underrepresentation was a matter of "math," he said, not "malice."[2]

Five years later, the same presumption that Graham pushed away—that the absence of women among startup founders must be a reflection of discrimination—irked Michael Arrington, the founder of TechCrunch. In an August 2010 post titled "Too Few Women in Tech? Stop Blaming the Men," Arrington addressed the *Wall Street Journal*'s recent criticism of Y Combinator for having funded only fourteen female founders to date.[3] Arrington suggested that the problem was not the acceptance rate but the paucity of women who wanted to be entrepreneurs. He speculated that, if anything, YC probably had a far higher rate of acceptance for female applicants than for male.

> I'm going to tell it like it is. And what it is is this: statistically speaking women have a huge advantage as entrepreneurs, because the press is dying to write about them, and venture capitalists are dying to fund them. Just so no one will point the accusing finger of discrimination at them.[4]

An outside observer watching the way YC went about selecting the teams for the summer 2011 batch would not see any evidence that female entrepreneurs enjoyed a special advantage nor were they openly devalued for being female. Paul Graham and the other partners with technical backgrounds were ideologically committed to a gender- and race-blind view of merit. Yin Yin Wu was a standout hacker, period. She did not come in under an affirmative action initiative. (Wu's relatively late start in hacking showed that there were side paths—albeit unusual ones, like computational chemistry—that could lead women into computer science after middle school.)

Graham says that, at the seed stage, investors are backing the founders more than the ideas, and he believes women are generally better judges of character than men. He says he himself possesses nothing more than an "average ability as a judge of character" and he credits Livingston, his wife, for applying an ability to size a person up quickly—what he calls her "perfect pitch for character"—to YC's investment decisions:

> Traditional VC funds have something of the atmosphere of a frat house. A grown-up version, which I suppose you might call a men's club. YC doesn't feel like that, and I think it could be a competitive advantage for us. There are certain ways men get stupid when they're together with just other men. You want to shoot from the hip. You don't want to seem weak. I think having a culture that's more balanced between male and female may make the atmosphere at YC more thoughtful.

✦

Before the summer batch applications had come in, Jessica Livingston had published on her personal blog the long essay "What Stops Female Founders?" offering advice that she hoped might lead more women to consider starting a startup. She does not argue that all women should give startups serious consideration. She addresses twenty-five-year-olds in particular and draws lessons from her own experience, fifteen years earlier, as a twenty-five-year-old who had a BA in English and lived in New York City, holding down a boring job at a financial services firm. "I wish I had learned to program when I had the luxury of spare time," she says. "Now I'd tell myself: take a class or get a friend to teach you. Even if you aren't very good, it will make programming seem less foreign and terrifying."

She recommends learning about startups by reading and talking with those with experience, and finding a technical cofounder if you aren't technical yourself—advice she says she would have had difficulty following; as an English major, she did not know any programmers. She also advises not living beyond your means and slipping into debt; instead, she recommends giving up creature comforts, saving, and trying a startup on a part-time basis, perhaps on weekends initially, to see if it suits you.

Prospective founders also need to be prepared for rejection. "You are suddenly in a world where you get slapped around a lot, so if you take slaps personally, it is going to be distracting," she says. Prepare also for discrimination: "It's been true in the past and probably is still true to some extent that investors discriminate against women. Not necessarily consciously, but their models of the ideal founder are current successful founders, who are mostly men." She ends on a hopeful note. Fortunately, startups require less money than ever to get started. The principal obstacle, she asserts, is the same one that prevented her when she was twenty-five from joining a startup: failing to even consider a startup as a possibility.[5]

✦

The first YC batch in summer 2005 was completely male and the summer 2011 batch of 160 founders was almost completely so, too, but for the two exceptions. The numbers suggest that nothing had changed. But YC now drew founders who were not necessarily in their early or mid-twenties. The batch had grown to a size in which internal stratification by age could be seen. Founders tended to cluster by age when chatting with one another: mid-twenties and under, a separate group of those in their later twenties, and then those in their thirties (one founder was over forty). Among those in the older half of the group, there was also an affinity subgroup of those who had children, a group that certainly had not existed in the first batch of founders.

Depending on the distance between their family home and Mountain View, the fathers in the group got to see their families daily, seldom, or not at all. Hamilton Chan, the sole founder of Paperlinks, dashed down to Los Angeles on many weekends to be with his wife and three children. Kurt Mackey, of NowSpots, returned to Chicago a couple of times to see his wife and four children. Alexander Stigsen, of TightDB, did not get to see his wife and two young children; they remained in Denmark.

Jason McCay and Ben Wyrosdick, the cofounders of MongoHQ, and their wives decided to rent out their houses in Birmingham and bring the families to Mountain View. Between the two, they had five children who were five years old or younger. To reduce expenses, the two families decided to share one rental house. What was available and within their budget was a 1,600-square-foot house that was far smaller than either family had in Birmingham. This meant that the founders were living and working in the close quarters typical of the lean startup—their office was a remodeled toolshed that sat in their tiny yard—made even closer by the fact that their families were living in the same confined space, too.

Chris Steiner of Aisle50, whose home was in Chicago, also elected to come out with his wife and three-year-old son. His cofounder, Riley Scott, who was married and had a one-year-old, was already living in the Valley. The two families consolidated households, sharing a three-bedroom house in Los Gatos. Each family had one bedroom; a developer who was working for them occupied the third. Steiner's mother-in-law and two sisters and the developer's fiancée also spent time in the house. (Looking back on the summer, Steiner would say, "The stress levels on our families were uncomfortably high—not something we'd like to repeat.")

The MongoHQ and Aisle50 founders represented the older end of the spectrum, where there was a greater likelihood that domestic considerations would complicate the decision to start a startup and apply to YC. At the younger end were founders like Yin Yin Wu, whose pre-YC baseline of comparison for creature comforts—her Stanford dorm suite shared with three others—made the tiny apartment of her own that she moved into at the beginning of the summer seem positively spacious. She says she was glad that she had not enjoyed a higher standard of living before coming to YC—it would have been harder to give up. "We're giving up cafeteria food for ramen. Not much different."

For Wu, the decision to come to YC has meant turning away from Facebook and others in the Valley, which offer generous salaries, stock options priced at pre-IPO levels, and other goodies.[6] She says her situation brings to mind the famous marshmallow experiment, done at Stanford's Bing Nursery School in the 1970s, which tested the ability of young children to exercise self-restraint. The subjects were told that if they resisted the temptation to eat one marshmallow while the researcher stepped out of the room, they would receive a second one. Those who did well in the test later thrived broadly as they reached adolescence and then adulthood. Wu says she is electing to forego the certain payoff of working as a software engineer at a large company in order to work for her own startup. She quips that she hopes "to get those two marshmallows later on."

✦

A year earlier, Jessica Mah went through YC as a member of the summer 2010 batch. She too was a newly graduated computer science major, in her case from Berkeley. But she was three years younger than Wu, only nineteen when she arrived in Mountain View. She had begun college at Bard College at Simon's Rock, the college in the Berkshires for teenagers who have no need to finish the last two years of high school. After transferring to Berkeley, she met Andy Su, another computer science major and a wunderkind himself: he was a year younger than Mah. The two began immediately working on side projects. In the spring of 2009, in their junior year, they decided to do a startup, inDinero, that would compete online against

Quicken's QuickBooks, offering financial management to small businesses, just as Mint.com had taken on Quicken's consumer product with an online service. They had a working prototype of inDinero in short order.

When Mah was asked later that year in a podcast interview why she chose to start an ambitious startup while she was still in college, she said that she did not understand "why people put too much effort into planning the future when it's right in front of them."[7]

Asked her thoughts about the rarity of women in computer science, she said she usually did not give the matter any thought, but a few days earlier it had occurred to her and she realized that she did not have a satisfying explanation. So she asked the women in Berkeley's computer science department what toys they had played with as children—and whether Barbies were among them—and the answers she received were consistent: no Barbies. Legos were the favorite toy.

She also noticed that most of the female computer scientists had either a mother or a father who was an engineer. This was true in her case, too: her father, an engineer, had given her Legos and science kits.

Mah and Su's startup received backing almost immediately in the form of a $35,000 check when inDinero was chosen as one of seven startups founded by college students who received a summer grant and office space from Lightspeed Venture Partners. They then received another cash grant while seniors when inDinero was selected as a winner in Berkeley's Venture Lab Competition.

In spring 2010, as graduation neared, the two founders applied both to YC and to TechStars in Boulder; they landed finalist interviews at both, and they received word of acceptance into both on the same day. It was difficult to choose. Mah said:

> TechStars is really hands-on, really intimate. YC, it's more: We'll let you run your business, we're hands-off. If you want help, they'll give you help. I thought both approaches were great. It's like going to a state school, where they're completely hands-off, versus private school, where they'll try to do much more to ensure that each and every person succeeds really well.[8]

Mah and Su decided to go with YC and its hands-off approach. That summer Mah was one of four women among the eighty-one founders in that batch.

The two inDinero cofounders recruited two software engineers as first hires even before the YC session began. Three weeks into the summer, Mah described the culture that she and Su were creating: "InDinero is actually like a family. We cook and clean for each other, treat each other like playful siblings, work as hard as you'd expect from a group of Asian immigrants."[9] Shortly after YC concluded, inDinero raised $1.2 million.

✦

"Where is the female Mark Zuckerberg?" asked *San Francisco* magazine in a cover story that ran in late 2011. "For the first time in startup history," the article asserted, "girl wonders actually have an edge over the boys." Leah Busque, founder of TaskRabbit, was featured, and shorter profiles were provided of more than two dozen others, including Alexa Andrzejewski of Foodspotting, and Susan Feldman and Alison Pincus of One Kings Lane. Jessica Mah was included, too. The caption accompanying her picture said, "The stereotype: Only boys launch companies when they're still in college. The reality: Girls have dorms, too."[10]

The tone of the article, which begins by emphasizing the supposed "edge" that women founders in the Valley enjoy over men, changes markedly near the end, when it suggests women remain as behind men as ever. The author, E. B. Boyd, expresses her dismay as she watches Busque, the founder she's featured most prominently, step out of the chief executive role at TaskRabbit to make way for a male outsider that the board has hired to take her place. And Busque stepped down cheerfully. The female founders of Gilt Groupe, One Kings Lane, and Silver Tail had all done something similar. Boyd was told that it makes sense to have inexperienced founders step to the side and learn from a more experienced executive. But she wonders why it is that someone like twenty-six-year-old Aaron Levie, founder and still chief executive of Box, an online storage service like Dropbox, can brush aside any suggestion that he step aside. "I've always wanted to build companies," Boyd quotes Levie. "The best way to keep me super excited is to keep me in the driver's seat."

Why do the boys get to stay in the driver's seat and the girls do not?

✦

It is a few weeks since the "Where Is the Female Mark Zuckerberg?" article appeared. When Paul Graham sits down in the New York studio of Bloomberg TV for an interview, he finds himself fielding a barrage of questions from reporter Emily Chang about what should be done to increase the numbers of women and minorities in tech startups.[11] He begins by saying that the makeup of a YC batch reflects that of the applicant pool. To get a picture of what the pool looks like, he suggests that Chang do a Google search for a tech conference, such as one for Rails programmers, and take a look at photos of the audience. "You'll see a bunch of white and Asian dudes, you know?" he says. "*That's* the pool of programmers."

Graham does see change in the startup world, with the appearance of some startups that "are *not* so much about straight technology as they used to be. Now maybe it's the Gilt Groupe, right?" This could be expected to produce changes in the demographics of founders, so "the problem could be improving somewhat."

Chang has a proposal for YC when it chooses founders to fund: adopting Eric Ries's suggestion that applications be submitted blindly, without any information about name, gender, race, or background. "That might lead to a different outcome," she says. "Would that work for you guys?"

"You have to talk to them in person 'cause it's so much higher bandwidth," Graham explains. How would blind interviews work? he wonders aloud. "If we tried to interview people behind a screen—that would seem pretty bizarre."

Graham returns to the point he has made for years, that the pool of startup founders—at least, the hackers that YC invests in—are playing with computers when they are thirteen years old. "If you want to fix the problem, that's what you have to change."

5

CRAZY BUT NORMAL

"I felt like an immigrant from Eastern Europe arriving in America in 1900," said Paul Graham, recalling the shock he experienced when he first moved to Silicon Valley in 1998. "Everyone was so cheerful and healthy and rich. It seemed a new and improved world."[1]

Graham worked at Yahoo—briefly—after the sale of Viaweb in 1998, and then moved back to Boston. After YC was started in 2005 and alternated between Boston in the summer and Mountain View in the winter, he saw that the two places were not equally congenial to startups; Silicon Valley had no peer.

American pundits often claim that Americans, as a group, are more entrepreneurial than any other country's general population. Graham emphatically rejected this. In his opinion, what other countries lacked was not entrepreneurial drive but a concentration of founders who had taken the plunge and showed others it was possible. He wrote in 2006, "Some say Europeans are less energetic, but I don't believe it. I think the problem with Europe is not that they lack balls, but that they lack examples."[2] European hackers simply did not have the opportunity to meet startup founders. This had nothing to do with culture or national character. It was a matter of geography. No place, inside or outside the United States, had as conspicuous a concentration of startup founders as did the Valley. "Stanford students are more entrepreneurial than Yale students," Graham observed, "but not because of some difference in their characters; the Yale students just have fewer examples."[3]

✦

When Graham and his three partners—Livingston, Morris, and Blackwell—looked over applications in late 2006 for the winter 2007 batch that would start in January, they invited a two-founder team from the UK to come over for an interview. While students at Oxford University, Kulveer Taggar and Harj Taggar, cousins who were twenty-one and nineteen, respectively, had founded Boso, a classifieds Web site for college students who wanted to sell stuff (Boso: "Buy Online, Sell Online"). The site initially served only Oxford; then the two attempted to open sites at other campuses in the UK.[4]

Kulveer was president of the Oxford Entrepreneurs Society and was drawn to the idea of startup life as a career path. But there weren't others who could have showed him that it was actually feasible and not just a daydream. At graduation, he accepted a position with Deutsche Bank. During the day, he was a banker trainee in the City; in the evenings, he worked on Boso. In the meantime, Harj was finishing his degree in law, spending less time on his classes and more time on Boso.

Every fall, Oxford's Saïd Business School hosts an annual conference, Silicon Valley Comes to Oxford. In 2005, the guest speakers included Chris Sacca, a Google executive, and Evan Williams, a cofounder of Blogger (Twitter came later). Kulveer had the chance to have drinks with the speakers and they encouraged him to quit banking and commit to startup life.

It was not easy to walk away from the bank. Kulveer had finished his initial training period and now was earning a nice income. He drafted a resignation letter, then discarded it. He tried again, and abandoned that, too. Finally, he presented his supervisor with the letter—and was pleasantly surprised to receive the blessing of his boss, who also offered to invest, giving him a check for £16,000 (about $30,000).

In early 2006, Kulveer became a full-time founder. When Harj graduated in June, he too was free to work full-time on Boso. The two young founders managed by August 2006 to raise an impressive £160,000 (about $300,000). They knew that capital was a necessary, but not sufficient, condition for making a go of Boso. They needed different surroundings, a place where experienced founders who could provide guidance were plentiful. Kulveer flew to San Francisco to talk with Evan Williams; he also met Max Levchin, a cofounder of PayPal, and Naval Ravikant, a serial founder. When he returned home, he told Harj that the conversations he had had with those people were incomparably more useful than those with people in London. That spurred the two to think about relocation to Silicon Valley.

One day, on a whim, Harj did a Google search for "mistakes startups make." The first link on the results page pointed to an article titled "The 18 Mistakes That Kill Startups," written by someone he had not heard of: Paul Graham.[5] Harj went down the list of mistakes one by one, and recognized many from his and Kulveer's own experiences. He sent the link to Kulveer, who wrote back, "Why don't we just move out there then?"[6]

They planned on coming over on their own. It did not seem like Y Combinator would be interested in them: it had funded three batches to date and seemed to be set up for founders who were hackers and American—they were neither.[7] But the two decided to invest the time to fill out the application for YC's winter 2007 batch—and became the first company funded by YC from outside the United States. Jessica Livingston, later reminiscing about that interview, said, "They were the only people we've ever funded who were not technical. *But!* They were teaching themselves to program! I heard that and I was like, 'You're business guys teaching yourself to program? I don't care what your idea is, you're in'—I loved them!"[8]

When the two Taggars arrived and settled in, Harj marveled at the friendliness of everyone they met. They got an apartment in San Francisco and Evan Williams offered them desks within his own space at Obvious Corporation—soon to evolve into Twitter—that was only a five-minute commute away. No more three hours of daily commuting as they had done in London.

Most exciting of all was their discovery that the people that they met in the Valley seemed to regard launching a startup as a perfectly normal thing to do. Here, "people seriously live and breathe startups, it's literally like a whole way of life unto itself," Harj wrote. In his first week, he went to a party and was surprised by what he saw:

> It was strange seeing all the consultants being almost embarrassed to say what they did—they went through the typical process of forming rationalizations as to why it was the best thing for them to be doing right now. I found it slightly surreal. My experience has always been of feeling like an outcast/eccentric/weirdo for wanting to start my own company straight out of university and I've often had to make arguments for why it's the best thing you can do if you want to invest in yourself. Being surrounded by people who see that as the norm has taken a little getting used to.[9]

Harj also quickly came to the realization that he and Kulveer were not as precocious as they had been led to believe back in the UK. There, the government had made much of their youth and had showcased them as model founders. In the second week at YC, the two were invited to dinner with the Reddit and Kiko founders from YC's first batch, along with some other founders. As Harj looked around the table, he saw founders who were just as young as he was, yet they all seemed to be working on their second or third startup. "And there I was thinking I'd done well to have turned down a safe career in law to work on my company with one round of funding. Compared to these guys I was pretty much at the bottom rung of the ladder."[10]

Contemplating the long odds of success would lead to hopelessness, Harj wrote on his blog. Even if founders happened to hit upon an idea with some promise, the startup would soon face competition from a giant like Google or Microsoft. Why bother? "In a way you're kind of crazy to even try and build a startup," Harj said. "Well, the reason I'm loving it out here is because I'm surrounded by crazy people—people who know they're competing against the odds but they do it anyway because it's what they want to do."[11]

After consulting Graham, the Taggar cousins decided to let go of Boso and the idea of classifieds for students and instead address a problem that they had stumbled upon: posting lots of listings on eBay was a cumbersome process. The two decided to use their time at YC to build a new product: software that would allow sellers to easily upload and manage eBay listings. Harj had no more than a beginner's acquaintance with coding. Now he had to "suck it in" and learn programming in a serious way, which reminded him of cramming for final exams.[12] It helped to be a member of the Y Combinator family. A neighbor, Srini Panguluri, who was a true hacker—Stanford computer science graduate, former Oracle employee, and YC alumnus working on a startup, YouOS—offered to serve as a tutor. Panguluri ended up working full-time that summer with the Taggars, before returning to YouOS.[13] With his help, the Taggars converted the new idea for an eBay tool into a rudimentary product that could be shown when founders presented prototypes of their products to their batch. The Taggars had a new name for the company—Auctomatic—as well as the new software to show, and were pleased that day to be voted by their peers as one of the three most promising companies in the batch.[14]

Later, at a party that drew many Facebook people, including Mark Zuckerberg, Harj marveled at how young its executives were and how leadership in the startup world was not based on age. He had taken the first steps down a career path in law, where pay and rank were determined by the number of years a person had worked for a firm and not by talent or ability or accomplishments. In the Valley, a different system was in place, one that allowed a Zuckerberg to run a company worth a billion dollars (this was early 2007) and, by all appearances, run it well.

During their YC residence, from January through March 2007, the Taggars sustained the feeling of exhilaration they had experienced when they had first arrived. After the session was over, Harj compared the experience with what he had found when he'd first arrived at Oxford and made the uncomfortable discovery that he was no longer the smartest student. So too at Y Combinator, he had been surrounded by people who were as smart or smarter, forcing him to work harder than he had before. This was the environment that he and Kulveer thought was ideal, the opposite of "big fish in a small pond." In the UK, he realized, he and his cousin had received attention in national media just because of the novelty of young founders starting a startup. They'd received the attention before they had actually accomplished anything of note. At Y Combinator, they were just one of many young teams of founders.[15]

Among the highlights of the experience was the exposure at the weekly dinners to accomplished founders like Paul Graham, Evan Williams, and Paul Buchheit, a very early employee of Google, who had created Gmail. Being in such company, Harj wrote,

> it suddenly dawns on you that there's actually a chance you can succeed despite stupid odds. Logically speaking you can argue that I'm looking at it from a skewed data point—for each of these success stories there's a hundred failures, you might say. And you might well be right. The point is that sometimes you just need to forget what logic tells you and just go after something because you want it that bad—YC gives you the belief to do that by placing you in the perfect environment.[16]

The Taggars would soon have reason to be glad that they had ignored logic. They were able to keep Auctomatic going with investments from more angels, including Buchheit and Chris Sacca. And with Graham's introduction, they added as a third cofounder a genuine hacker, Patrick Collison, an MIT student from Limerick, Ireland, who was only nineteen and had applied to the summer 2007 YC session as a single founder. Graham steered him to Auctomatic instead of having him go through the summer batch. Then Patrick's younger brother John, who still had two years of high school to finish, also joined Auctomatic, quickly learning how to code on the job.

Y Combinator's motto was "Make Something People Want," and the four Auctomatic cofounders did so. EBay's power sellers were quite taken with Auctomatic's inventory management tool, which was free. It supported eBay's international sites that seemed obscure to Auctomatic's large competitors and found greatest acceptance among eBay sellers who were outside of the United States.[17]

As it added users, Auctomatic caught the attention of prospective acquirers. Less than a year after completing YC, Auctomatic, then consisting of five people who lived in a two-bedroom apartment to save on rent, was presented with an acquisition offer by a Canadian company, Live Current Media, based in Vancouver. While they were considering the offer, two other companies jumped in, expressing an interest in acquiring the company.[18] When the bidding ended, Live Current Media prevailed. The $5 million acquisition price consisted of $2 million cash with the remainder paid in Live Current Media stock.

The news gave Oxford University's press relations staff flutters. "Graduate Entrepreneurs Sell Business for Millions" was the headline of a university press release when the deal was completed in May 2008.[19] Referring to "millions" was a bit misleading. The greater portion of the deal consisted of unvested stock that the cofounders had to acquire over time by working for Live Current Media in Vancouver. The cash proceeds had to be split with the investors and debts had to be paid off. They were certainly better off than they had been, but their financial position did not resemble that of Paul Graham's after Viaweb was sold in 1998 for $50 million.

Harj worked at Live Current Media in Vancouver for a little more than a year, then moved back to San Francisco in August 2009, planning to do another startup.[20] For the first one, his motive had been mercenary: "I'd never had much money growing up and the desire to be rich was hardwired in me since as far [back] as I could remember." But having seen firsthand the difficulties that startups face, he could no longer justify the effort as a mercenary enterprise. "My brain just won't let me delude myself on the chances of that outcome like it used to."[21] Now he only wanted to work on a startup that attempted something intrinsically interesting.

He came down to Mountain View to speak with Graham and Livingston about ideas and then went back to the UK to spend time with family. In December, he received an e-mail from Graham asking if he would be willing to take part in an experiment that Graham wanted to try: having someone besides himself hold office hours with YC founders. Through nine batches, Graham had been the sole adviser. Livingston handled the occasional intracompany conflict among founders, as well as legal issues—it was she whom the Taggars credited as being instrumental in arranging for their visas[22]—but she did not weigh in on product questions or other strategic matters. The other two YC partners—Robert Morris, at MIT, and Trevor Blackwell, at Anybots—were busy with day jobs. Graham said the winter 2010 batch would be three times as large as Harj and Kulveer's winter 2007 batch and he needed help. Would Harj be willing to see how well it would work? After flying back, getting a look at the winter batch, and trying the idea on for size, Harj accepted.[23]

The temporary position became a permanent one later that year when Harj Taggar was named as a YC partner.[24] Deciding to stay at Y Combinator was more difficult for Harj than the decision to come out and serve temporarily. "There's so much luck involved with startups you increase your odds of success by swinging the bat multiple times. Each time you do something that isn't swinging the bat, you theoretically decrease your odds of success," he said. But the thought of missing out on YC's audacious experiment, funding so many startups at a time, was more painful than the thought of delaying the start of his next startup.[25]

At the same time that Harj Taggar was formally welcomed to the ranks of YC partners, so too was Paul Buchheit. Graham noted that Buchheit "was responsible for three of the best things Google has done": developing Gmail; developing the prototype of Google's advertising system; and coming up with Google's guiding mantra, "Don't Be Evil." After leaving Google, he had gone on to cofound a startup, FriendFeed, which the previous year had become Facebook's largest acquisition to date. Graham also gave him the highest praise he gave anyone: "one of the world's best hackers."

The software technology that Graham had cut his teeth on at Viaweb in the 1990s was only a distant relation to the technology that hackers were working with now. New partners, however, gave YC a broader generational base of technical experience: Harj Taggar was about half the age of Graham; Buchheit's age fit in between the two. In June 2011, YC added four part-time partners, all of whom were YC alums: Garry Tan, a cofounder of Posterous, who became the designer in residence at YC, and three founders who had been in the first YC batch in Cambridge in 2005: Sam Altman, the cofounder of Loopt, and Justin Kan and Emmett Shear, the two Yale students whose startup career had begun with Kiko, the online calendar site.[26] Near the end of the summer batch, Tan became a full-time partner and a new face appeared: Aaron Iba, a cofounder of AppJet, a YC-backed startup that had been acquired by Google.[27]

Graham had managed to launch YC with only one experienced founder to serve as adviser—himself. It was a good hack. But now YC was operating at a scale in which it had the resources to make these strategic additions.

✦

To accept the role of partner, Harj Taggar had to suppress the urge to swing the bat again and try doing another startup. But the Collison brothers were not so constrained.[28] After Auctomatic was sold, Patrick worked for about a year at Live Current Media and then returned to MIT in fall 2009 for his second year of school. John, after returning to Ireland to finish high school, arrived for his first year at Harvard, joining Patrick in Cambridge. Before the fall semester was over, the two had decided to see whether they could build a product that would simplify the process for Web developers who are setting up software for receiving credit card payments. The Collisons wanted to make setup simple, painless, and instant—everything that PayPal was not.

Both of their universities had a monthlong semester break in January and the two decided to leave the dreary Cambridge winter and head to Buenos Aires to start coding. It was an inexpensive place to stay, filled with WiFi hot spots and organized on what Patrick Collison described as a "hacker-friendly" schedule: restaurants stayed open until two a.m. and bars until five a.m. or later. Two weeks after they arrived, they had a functioning service and their first user, Ross Boucher, a cofounder of 280 North, a YC-funded startup (winter 2008).

The two returned to school for their spring semesters, but were itching to work full-time on their little payments company. In June 2010, they went out to Palo Alto, thinking they would devote a summer to determining whether they had an idea worthy of pursuing. "We weren't really sure how big a problem this was," Patrick Collison would say later. "Maybe there'd be a nice niche for a small, little, extremely developer-focused payments system, but perhaps that could never be something all that significant."

When they arrived in Palo Alto, they first paid a call to Paul Graham, who immediately had Y Combinator invest. The company would be called Stripe and became a member of YC's summer 2009 batch. The Collisons drew encouragement from the way that the friends who tried out their service asked if they could share it with their friends, and those friends asked, too. "That was kind of surprising to us," Patrick Collison said. "It's a payments system, not a social network! It's not immediately obvious that you'd have any kind of virality whatsoever."

The incumbent payments services were so painful to use, however, that developers latched onto Stripe as soon as they heard about it. The Collisons had taken on a problem that was indeed a large one—one that had been ignored by other startup founders searching for ideas. Graham later praised Stripe for taking up a great startup idea that was "lying around unexploited right under our noses."[29] The opportunity was available because others were limited by their aversion to extreme difficulties:

> Probably no one who applied to Y Combinator to work on a recipe site began by asking 'Should we fix payments, or build a recipe site?' and chose the recipe site. Though the idea of fixing payments was right there in plain sight, they never saw it, because their unconscious mind shrank from the complications involved. You'd have to make deals with banks. How do you do that? Plus you're moving money, so you're going to have to deal with fraud, and people trying to break into your servers. Plus there are probably all sorts of regulations to comply with. It's a lot more intimidating to start a startup like this than a recipe site.

When Patrick Collison was asked what he thought about Graham's point about the intimidating nature of a big problem, Collison politely dissented, arguing that what should be emphasized was that addressing a hard-to-solve problem is actually not as hard as everyone thinks: founders and employees alike are inspired to work harder than if they were taking on the mundane.

Midsummer in 2010, the Collisons met with Peter Thiel, the cofounder of PayPal, who expressed some concern that Stripe was a "two-stage" rocket. "There's all these fantastic things you can do at stage two, but it depends on a really good execution of stage one," Collison recalled Thiel saying. Nonetheless, Thiel offered to invest.

At the end of the summer, the Collisons completed a $2 million round raised from Thiel, Sequoia Capital, Andreessen Horowitz, SV Angel, Elon Musk—another PayPal founder—and a few others. The Collisons did not return to school in the fall.

Then, barely eighteen months after the two had begun working full-time on Stripe, the brothers—still only twenty-three and twenty-one—raised an $18 million round, led by Sequoia, at what was reported to be a $100 million valuation.[30]

✦

The only remaining member of Auctomatic's original team of four who had not come back to YC or started another startup was Kulveer Taggar. But he too ultimately succumbed to the allure of startup life—he's been accepted into YC's summer 2011 batch.

He now has two technical cofounders: Srini Panguluri, the YC alum who had worked closely as his and Harj's programming tutor in the early days of Auctomatic, and Omar Seyal, a close friend of Panguluri's at Stanford and a fellow computer science major. The three want to tackle a big idea. Their startup, Tagstand, will promote a new technology, NFC (near-field communication), that is coming to smartphones, enabling many things to be done with a tap of the phone.

The year before, when Kulveer complained about his unproductive state, Harj had reminded him that one's environment is critically important to one's productivity. Now Kulveer was back at YC, surrounded once again by ambitious, optimistic, and hardworking peers. Harj was right: surroundings matter.[31]

6

UNSEXY

In Y Combinator's main hall, the distinctive all-orange wall with spiky soundproofing panels, a YC signature from its beginning, is in place at the far end. The hall has been much enlarged. When YC first moved into its Mountain View space, it needed room for only that thirty-foot-long table to seat the founders of the eight startups comprising the winter 2006 batch. One wall has been pushed back three times and now the hall holds six of these long tables to accommodate the 160 founders in the summer 2011 batch.

On Tuesdays, by late afternoon, the aroma of chili or spaghetti or a Chef's Surprise of uncertain composition wafts from the Crock-Pots. Founders begin arriving earlier, setting their laptops on the tables and working while waiting for their office hours appointments. The hackers on most of the teams are writing code, working on a first prototype, an existing product that needs new features, or a product that needs to be rewritten from scratch.

Some teams are beset by doubts about the idea they proposed in their YC application. The more they scrutinize the idea, the less promising it seems. The essential nature of a startup is not that it is a new business but that it is a new business set up to grow very quickly—in Paul Graham's phrasing, it must be "designed for scale."[1] Thus, a programmer who decides to open a consulting business to help clients build Web sites is starting a small business, not a startup. But a company that sets out to build software that automates the creation of those Web sites, a product that has the potential of being distributed quickly and broadly on a mass scale, fits the definition of a genuine startup. Or at least it does so at first glance. One of the companies in the summer batch will take up this idea and try it out. Its founders will learn, however, that if you write software that automates Web site creation for small businesses, but the service must be sold in person, door-to-door, then it is a business that lacks that defining essence of a startup, the ability to scale quickly.

One of the teams that is paralyzed with self-doubt is the Kalvins—Kalvin Wang, Jason Shen, and Randy Pang, the three recent graduates of Stanford and Berkeley who had charmed the YC partners with their retro idea to publish printed photo books. After the first dinner's formal program is over, the Kalvins ask Graham to sit down at their table and help them.

When they had spoken weeks earlier via Skype with Harj Taggar about their original idea, of photos emailed to users' inboxes, he had pressed them to explain how they would attract users. Unable to supply a satisfying answer, they had later come up with the idea for printed photo books. But this did not answer that question about where they would get users.

At this point in their recital of how they had come to waver and wonder whether photo books is the idea they should pursue, Graham excuses himself for a moment to go over to his laptop and look up what he had written in his notes after their interview. When he returns, he reports that he had written the following: "Insanely energetic founders. Fund for the new idea." So Graham is not going to be the one who encourages them to pursue photo books, either.

Graham tells the Kalvins, "Here's how to generate new ideas. Three things. One: founders are target users. Two: not many people could build it, but founders are among them. Three: few people realize it is a big deal."

During their YC interview a month earlier, Graham had brought up the Altair personal computer that was the improbable basis for Microsoft's start. Graham brings it up again—he loves this particular parable about how a small idea grew into a diversified powerhouse of a business. Microsoft had started with the BASIC programming language for one brand of microcomputer. Then it added variations of BASIC for many brands of microcomputers as they proliferated. Then other programming languages. Then operating systems. Then software applications. "Then they went public!" he says excitedly. "Four hops!" The sequence had begun with Bill Gates's recognition of the opportunity to solve a problem: endowing the first personal computer with the general-purpose software that would allow the machine to be programmed.

"Ask yourself: 'What do I wish someone would start a startup to do for me?' " says Graham. "The next best thing: something for someone else that you know is a problem."

The Kalvins have tried to answer these questions on their own but their efforts have not yielded much so far. "We feel like we have some problem areas that we're really passionate about. But nothing comes up that's like: this will actually solve the problem better than existing alternatives," says Jason Shen.

"Do you ever, in your work, say, 'Boy, I wish somebody would just—*blank*'?" asks Graham. He tries another tack, asking about areas in which they are "domain experts," the way Google's founders were experts in search.

"We all sort of collected our strengths and things we're good at," Shen says. "It's the most random things."

Graham offers another inspiring tale about a startup that addressed an unmet need: Apple, which was started by Steve Wozniak because he wanted his own computer. "He couldn't afford the components. So he designed computers on paper. And then DRAMs came along—chips became just cheap enough that he could build a computer." When Steve Jobs saw what Wozniak had built, he suggested that they sell it to other people.

Looking back, the need seems obvious. Graham suggests that they ask themselves: "What will people say in the future was an unmet need today?" He says, "My Mac Air takes like a minute to boot up." He raises his voice to a high pitch to show his exasperation. "It's *ridiculous*!" His voice drops back down. "I'm telling you, just in a few years from now, people are going to say, 'Can you believe people would just sit there and wait for like three minutes for a computer to boot up?' People won't even know what booting up is, right?" He imagines a future exchange between youngster and oldster:

"What do you mean? What is 'booting'?"

"Booting is this thing that happened between when you turned it on and you could use it."

"Really? There used to be a delay?"

Graham conveys the sense that good ideas are plentiful, waiting for someone to come along and pluck them off the ground. "There's a bunch of things like that," he says.

This is as far as he can take them, however, and he gets up. The expressions on the faces of the Kalvins do not indicate they see a bunch of things like that.

✦

BrandonB is the temporary YC name of another startup that is unsure about what it should work on. When Brandon Ballinger and his cofounder applied to YC's summer batch, Ballinger had the educational and professional background of a prime candidate. He had earned a degree in computer science from the University of Washington and then gone directly to work at Google. He had just resigned from the search company after almost five years there.

When BrandonB's interview time arrived, Ballinger walked into the room alone, without his cofounder.

"There's only one of you," Graham said as his greeting.

"Yeah." Ballinger's cofounder had suddenly decided to abandon the startup. "That's the bad news that happened at twelve thirty a.m. last night," Ballinger explained.

Graham reassured him that he was not the first YC founder to have been stood up. "It's the first time *I've* been stood up," Ballinger said. He laughed, and so too did the YC partners, seeing that he was bearing up despite the setback.

Ballinger explained that he had changed the idea he had sent in with his application. His new idea was to build a mobile app that would show all of the local events in one's vicinity, providing more complete listings than existing event services did. He had reviewed the computer science literature on "event extraction"—using software to visit Web sites and pull out information about upcoming events—and described the process ("a Hidden Markov model which pipes into a CFG parser, which pipes into a semantic analyzer. It's actually pretty similar to the way a speech recognizer works").

A startup devoted to hyperlocal events struck the partners as a terrible idea—how much demand would there be for a comprehensive listing of neighborhood bars' happy hours? Worse, Ballinger was now a single founder and YC generally did not fund single founders. But the group saw promise in him nonetheless and welcomed him into the summer batch.

YC is now under way and Brandon Ballinger comes in for office hours with Harj Taggar. Ballinger has good news to report: he has recruited a cofounder, Jason Tan. They had gotten to know each other when Tan had taken a computer science class at UW for which Ballinger had been the teaching assistant. After graduation, Tan had worked at three startups. He is moving from Seattle to San Francisco to join Ballinger and is on the highway this very day, with all of his possessions in the car. When Ballinger walks into the small conference room, he has Tan on the line and turns on the speakerphone so Tan can participate in the conversation from the road.

Ballinger and Tan are reconsidering the original idea to do local events—it is one of five ideas that they are weighing. The first idea is spam detection for clearing out advertisements in public discussion forums. Ballinger has spoken with the chief executive of one of YC's well-established startups, who told him that spam was a significant problem on his Web site's chat forums and that he "would totally want somebody to come solve this for him." Ballinger says he is excited about this idea because he worked on detecting keyword spam at Google, and Tan has worked on related problems, too. Ballinger is concerned, though, that fewer than a thousand Web sites host discussion forums whose scale is large enough to need the service. This would seem to limit the potential size of the startup.

Another idea they are considering is medical resident scheduling. "Do you know how it works? OK, at the beginning of the year, the head resident puts together a schedule of all the rotations medical residents do. Each one has a specialty—"

"How did you arrive at *that* idea?" Taggar asks.

"We know a lot of doctors. Jason has been on the phone with like ten to fifteen different doctors. I heard about it 'cause my doctor friends are always complaining that this is a huge issue—"

Taggar attempts to steer the conversation away from this idea and back to the first one. "The spam thing. You guys had worked on relevant areas in the past. Of the other ideas that you have, which are the ones you think leverage your strengths as a team? The doctor thing—you're essentially going to be relying on the words of friends to try to figure out what the right product decisions should be. It's kind of a crappy place to be."

When Ballinger tries again to make the case for the original idea of working on local events, Taggar is not encouraging. He observes that many startups are addressing this.

"Yeah, it's a crowded space," concedes Ballinger.

"*We* were actually skeptical—like when you came to the interview—"

"I know. Paul Graham's words were 'Jesus Christ!' "

Taggar laughs. "Well, it was like the hundredth one that we'd seen. The only thing that was impressive was you just seem to have thought it through more than other people had."

"I think the advantage that we have there is we can crawl the Web and get *everything* in your neighborhood, as opposed to the 10 percent that you see on Zvents or Eventful or whatever."

"What are the other ideas?" asks Taggar.

"So, the other ideas are subscription-based groceries. Do you cook a lot?"

"No, not a huge amount, actually. What I want is subscription-based decent meals."

"Do you want them precooked?"

"Yeah. I don't know if it's a particularly good startup idea, though. Go on with the subscription-based grocery stuff—how would that work?"

"A lot of the people we've talked to would like to cook more. To sauté something takes like fifteen minutes. But to go to the store, get the ingredients, find a recipe—it ends up taking like an hour, an hour and a half. That's why I don't cook. So the idea is, we'd just send you the ingredients necessary for a meal. At the beginning of the week, you'd say how many meals you wanted. And then you'd get the joy of cooking without all the fuss of figuring out where to get stuff and how much to get."

Taggar seems doubtful. "It has the hallmarks of an idea that is a potential nice-to-have. But it's not solving some burning, burning need. How many people do you know who are like, 'Man, the biggest problem in my life—' " He stops midsentence. "I mean, it doesn't have to be the *biggest* problem. But constantly complaining about wishing they had a subscription-based grocery service? Is there one more idea?"

Ballinger describes a "social travel" Web site: "This is like a virtual corkboard that you use to organize your trip." After the trip, the Web page would serve as the place to preserve memories of the experience.

This sounds like what an earlier YC-funded startup, The Fridge, had attempted. He steps back from Ballinger's list of ideas. "What kinds of things do you both like doing? What sorts of things do you enjoy working on?"

"I have a bias for things that are stuff that my friends would be using regularly, that's kind of cool," Ballinger said. "Or stuff that is very technically difficult. The second bias is probably a bad thing in a startup—"

"No! Technically difficult things are good, actually. Well, they're good if the technical difficulties are related to solving the problem. Building ten difficult things creates a barrier to entry; it's actually good." Which of the five ideas, Taggar asks, are they most excited about?

"In terms of what would use our skills, I feel like the event thing and the antispam thing would probably be the ones we're best at."

"The *best* kind of thing to work on—and I appreciate this is going to be somewhat abstract or higher-level advice—the thing you want to work on is, there's this need that's really clear and you can just launch some shitty site and people just start using it." Initially, the number of users is less important than the intensity of their attachment. Later, founders can usually figure out ways to expand the service.

Ballinger says that he and Tan asked about sixty friends to rate the two ideas they had for consumers: what he called "Netflix for Groceries" and

"Events Around Me." The survey results were bimodal: the friends either liked the ideas a lot or not at all.

Friends can mislead, Taggar says. His advice: address what businesses need, not what consumers say they would use. "With consumer stuff, there's more kind of randomness involved. It's not quite clear what people are going to use." He offers the example of Foursquare. Who would have predicted its success? "You can construct these narratives afterward about why they won. But it's hard to predict. So I generally think it's cool to work on things that make money and have some clear initial need." That's why he wants to know more about their spam detection idea. How would it work?

Ballinger explains, "Every time you have a comment, you send it to us as well as some metadata, like an IP address. We tell you whether it's spam or not. We'll give you zero to one, how spammy is it."

Another thought occurs to Taggar about how to assess startup ideas, one used by Sequoia Capital. "One of the big things they focus on is 'proxy for demand.' Which basically means when looking at some new idea, they want to see what people are doing at the moment? What kinds of crappy solutions are they hacking together at the moment." They'll ask founders who are building a product what their future users are doing right now. "If the answer is, 'No one is really doing it at the moment'—a lot of people think that's a good answer, but it's not. Because it means they're not desperate for it."

The spam detection idea is looking better. Ballinger says one company that they spoke with said it would be willing to pay annually the equivalent of "two engineers' salaries" for this service, because it was contemplating having two software engineers work on the spam problem full-time.

"That's pretty encouraging," says Taggar.

Ballinger and Tan are not yet ready to give up medical resident scheduling, however. "The adjacent thing is emergency room scheduling," says Ballinger. "Happens monthly. And there are a lot more ERs than residency programs."

Taggar still is not enthusiastic. "Of the ideas we've talked about so far, the spam idea sounds interesting to me because it's something you guys have a background in, it's something that someone has told you they would pay you for. It's unclear where you continue growing it, but it's one sort of company. It's a boring, unsexy company that makes money."

Ballinger is not dismayed to hear this. "Spam is pretty cool!" He laughs.

"See, that's an advantage—that you don't think it's boring!" Taggar quickly reviews the other ideas they had proposed, demand for which was highly uncertain. "It's up to you guys. Essentially, it's way better to make the wrong decision than it is to make no decision at all. Obviously, the idea that you work on is important. But try and pick things where the cost of failure is low and the cost for you is basically time."

The conversation has lasted forty minutes and it's time to wrap up. As Taggar walks Ballinger out to the large hall, he adds a few more words of advice, anticipating that Ballinger and Tan will drift back to their ideas for consumers, like the local events Web site or ready-to-cook meals.

"You want to work on something where if you see it among your friends and family, they go out of their way to go and tell other people about it," Taggar says. "Because friends will use stuff because they're friends, right? But the true test is if they start bringing other people into it. And *that's* something that they won't do out of obligation to you."

✦

Harj Taggar was not demonstrative and he never raised his voice. But with his soft-spoken manner, he has conveyed his opinion that spam analysis is the idea that seems to have the most promise. He has done his best to warn the two cofounders away from the ideas that would be offered to consumers. But they are free to shop around, seeking guidance from the other YC partners. Once Jason Tan has arrived in San Francisco, he and Ballinger sign up for office hours with Sam Altman. Perhaps he will give them the encouragement to pursue one of the consumer ideas.

Ballinger and Tan begin by explaining to Altman that they have five ideas, two of which are business to business, but it is the remaining ones, the consumer ones, that most interest them. Tan tells Altman, "We really want to build a consumer business. We want to build something that our moms could use."

"Hold on," Altman says. "Let me bring in PG."

Paul Graham comes in. The cofounders again state, with enthusiasm, their preference. "We really want to do consumer. We're really excited by that stuff."

For weeks Graham has been telling them this would be a mistake and yet here they are, blithely talking as if they have never proposed this before. "Moments like these are why I'm glad we invested in sixty-four startups," he says, deadpan. "If you want to drive off a cliff, go ahead." He walks out of the room.

The founders decide not to do a product for their moms.

7

GENIUS

"Launch Fast" is Paul Graham's mantra. Move from the idea to a minimally functional product as quickly as possible. Only by getting a product into the hands of customers, even if the product is only a prototype, is it possible to know what customers want.[1] Launching fast is how to make something people want.

Judging by the advice that they are given, startup founders are not naturally inclined to launch fast. Startup gurus have devised different ways of saying the same thing: launch the product even when it is in a bare-bones state. Eric Ries speaks of the urgent need to introduce a "minimum viable product," or MVP.[2] Steve Blank speaks of a "minimum feature set"—as soon as that set of features is operational, the product is ready to be tested in the market.[3] Graham has his own term but it is a clunker: a "quantum of utility," which means, in his words, a product that would make the world "one incremental bit better." The most widely circulated version of the idea is Reid Hoffman's, who is credited with saying, "If you're not embarrassed by the first version of the product you've launched, you've launched too late."[4]

The first companies funded by Y Combinator, back in summer 2005, began with nothing other than their ideas, and they all received the same advice: finish a prototype as quickly as possible; launch; get feedback from users; improve the product and release again, iterating and gaining more users while keeping a constant eye on the calendar as Demo Day approaches.

Among the hundreds of YC companies that followed in succeeding batches, a few took on ambitious projects that would not be ready, even in minimal form, by Demo Day. Some took a very, very long time. The current YC record holder for the longest interval between funding and product launch is Clustrix, from the winter 2006 class: its database software required four years of work before it was ready for unveiling.[5]

To the companies in the summer 2011 batch that are starting with just an idea, a four-year development schedule is unimaginable. Even using the full three months before Demo Day to prepare the first rudimentary version of the product will seem unacceptably dilatory. The cadence of startup life is speeding up considerably. At the time of the YC kickoff meeting, some startups in the summer batch already have a sizable number of users. In a few cases, they have revenue. These fortunate ones do not have to contend with anxiety about whether their idea will fly. They don't have to use office hours to get YC partners' guidance about what idea to pursue. They do not agonize about whether they should drop one untested idea and substitute another.

Rap Genius stands out as one of the most fortunate ones. Its Web site, which provides explanations of rap music lyrics, has already attracted a large base of users that is growing at a vertiginous rate. YC has just begun, yet every founder in the batch can see that the Rap Geniuses will have an impressive story to tell on Demo Day.

Paul Graham is fond of a Silicon Valley saying: "You make what you measure."[6] The act of measuring some aspect of performance leads to caring about that one thing and lavishing your attention on it, which, in turn, leads to improving it. He suggests that startups set a specific weekly target for growth, which can be measured in terms of revenue or users or something else, but should be essential to the startup. Anything that is not directly related to that metric is to be pushed to the wayside.

Aim for 10 percent growth—per week, says Graham. He knows that ambitious founders do not like the feeling of having failed, even briefly. He writes in the YC User's Manual:

> If you miss one week, you'll be thinking all the time next week about how not to fail again. And that is why such targets work so well: they cause you to focus. There are always so many different things you could be doing in a startup. You could be writing some new feature that sounds like it might be good; you could be trying to get mentioned in the media; you could be trying to meet customers; and a hundred other things. But when you're trying to hit a specific growth target each week, you'll find you think really hard about what to spend your time on. Before spending three weeks on a new feature, you'll be thinking: is this going to help us make our numbers or not? Could we write a simpler version in a day and see how much users like it before risking three weeks? Before you spend a day talking to the press, you'll be asking yourself: how many actual users is this going to bring to our site? That's 'You make what you measure' in action.

This is advice that can be put to use only after a startup releases a product, a prospect that at the beginning of the summer seems dismayingly distant to some YC startups. Or, in the case of Rap Genius, the advice seems irrelevant because traffic to its site is growing without the founders having to act. Most of their new visitors come from Google, where they've searched for a swatch of rap lyrics and clicked on the top listing on Google's search results page: Rap Genius.

Rap Genius is also growing because its users spread the word among their friends, who do the same. This is the same natural way that Wikipedia grew: the editorial content of the site draws users, a portion of whom donate their time to add to the content, which, in turn, draws still more users. At Rap Genius, users contribute the site's content—the annotation of lyrics—receiving as compensation only an improved "Rap IQ" score, which is displayed by the contributor's screen name. For now, Rap Genius also resembles Wikipedia in that it too is a Web site free of advertising or subscription charges. When Rap Genius presents the growth graph to prospective investors at the end of the summer, it will be able to show growth only in users, not revenue. But having a lyrics site that has no advertising is of great help in building up its fan base, attracting users who are besieged at other lyrics sites with pop-up ads and ringtone offers. Rap Genius can also lay claim to being the most authoritative rap music lyrics site. It describes itself as "basically the internet version of the nerd-ass 'rap dictionary' dorm-mate you had in college."[7]

The Rap Geniuses do not have to worry about "Launch Fast" or the growth graph for Demo Day. They do not feel the same urgency to look ahead as their batchmates. Just about everything about Rap Genius's genesis places it apart from the rest of the summer group. Not only has it gotten the running start, its founders were able to get it going as a part-time project while keeping day jobs. And it is a service that allows them to spend their working hours immersed in the music they all love. In sum, they are blessed with good fortune in getting started.

✦

Three weeks after the kickoff meeting, two Rap Genius founders, Tom Lehman and Mahbod Moghadam, come in to speak with Harj Taggar for office hours. The third cofounder, Ilan Zechory, has not yet moved to California from New York City; he is present with the help of a speakerphone.

Lehman begins with a brief report: "Growth of traffic is very good right now," he says. Traffic for the most recent two weeks has increased 25 percent over that in the preceding two-week period.

Is all of that growth coming from search engines? Taggar asks. Most if it is. More than 80 percent of Rap Genius's visitors are sent over by a search engine.

"Is there any new stuff you want to show me?" asks Taggar.

"Uh, no."

"So what specific product stuff do you want to talk about?"

The answer takes a little while to emerge. Rap Genius's most devoted users have no place at Rap Genius's Web site to discuss lyrics and matters of shared interest, so on their own the users created a public group at Google Groups for this purpose. To Lehman, this development presents a vexing question: should Rap Genius create a discussion forum on its own site, keeping its users from straying elsewhere? Or would the users' discussions pose a distraction, pulling them away from annotating lyrics, the essential activity that built the site?

Taggar shows no indication that he sees discussion forums as the critical question facing Rap Genius. He rubs his face and speaks quietly and rapidly, coaxing the founders to think about some other things. "If we just take a step back from this stuff, if we have to try and guess, if we went a few years out—say, I told you, five years from now, this is like some huge, massive company that's returned from the future, right?—and I'm making you guess why you're a big company, what would your guesses be?"

"I think the social element is a big part of it—" Lehman begins.

"What do you become? Do you become the social network of people interested in lyrics?"

Mahbod Moghadam tries out another vision for Rap Genius in the future: "Hopefully, if we got really big, we'd get a lot of artists—there are all these tweeting rappers—it'd be nice to have Rap Genius rappers—"

Lehman has still another answer to the question about the site's future: "I'd say we become the best lyrics site," he says. "The lyrics site where, in addition to reading stuff—and also in addition to being able to annotate songs—you'd also have some kind of social experience with people who like the same music, whether it's debating a line or just hanging out or whatever."

"That's never going to be the majority of people, right?" Taggar says. His questions suggest a skepticism about lyrics sites, which do not offer the music itself; they only offer the lyrics. How appealing can it be to talk about music, without listening to it at the same time? Taggar is curious about Rap Genius's hard-core fans. What kind of things do they want at the Rap Genius site? Have the founders spent time speaking with them?

"Everyone wants an app," says Moghadam. "Everyone wants a Shazam that tells you the meaning of a line." He says he forwards the requests for an app to Lehman, the coder. The response from Lehman is always the same: "Thank you, Captain Obvious."

"What you basically want to figure out is some kind of vision or some strategy for what makes this a big, interesting company," Taggar says.

"The plan was always to cover all of lyrics, at least," Lehman says.

"Maybe more than lyrics—poetry or something. Rock would probably be the next thing." Rock music, however, constitutes a separate vertical market (often shortened to "vertical")—that is, one that has particular characteristics and its own separate set of customers.

"Of all the stuff we've talked about so far," Taggar says, "the climbing toward the verticals sounds like the most interesting thing. If you plan to raise investment, it will also help. The difficulty that you're going to face if you want to pitch investors—presumably you do—is convincing them that this is more than you stumble across a site that has rap lyrics. That you come across as having some formula for repeating what you manage. If you can build a rock version of this that also gets to traffic quickly, then, one, you convince them you're not a one-hit wonder, right? You have some idea of what you're doing and you're getting to some kind of scale—that will be interesting. And it would be useful for you guys to know, can you even expand? *Can* you go into other verticals?"

"I'm sure we can," says Moghadam. But he says Lehman is the one who is reluctant to add another genre of music, like rock. "He's always worried about 'imperial overstretch,'" Moghadam says. "Tom has learned the lessons of the fall of the Roman Empire. Whereas me, I'm like Nero—I'm fiddling, and I just want to see it burn. I'm ready for *Rock* Genius."

Lehman says, in his own defense, that Rap Genius enjoys "organic growth," which is a rare thing that he does not want to let go of.

"I understand that fear," says Taggar. "But the fact is, if you want to make this into something interesting—what you really want to do, at some point, is just have this nailed-down, repeatable formula and continue growing it."

Moghadam says Rap Genius does have some rock lyrics and the collection has been growing organically, too. "I'm not the only one who puts up rock songs anymore."

"That's encouraging!" says Taggar.

"They don't get a lot of traffic, though," says Lehman, who is not as enthusiastic about building out rock lyrics now.

The founders give Taggar a tour of Rap Genius's Weekly Rap IQ Leaderboard, which shows the top contributors to the site. Points are awarded for writing what editors deem to be a good explanation or comment. Those who earn the most points over the course of the week have their names and Rap IQ scores displayed prominently on the home page.

Taggar is moderately approving. "I believe this kind of thing will be an important part of your product. Tapping into people's vanity and desire to show they're experts on things." He thinks for a moment. "These people are influencers. If you can start identifying people who know their shit, and therefore, basically, influence other people's taste—that's valuable for artists, record labels, various people. It's kind of what makes Twitter interesting."

Lehman returns to Taggar's suggestion that Rap Genius add other genres. "The thing about doing the other verticals thing is that it's such a big project," he says. He's the only one of the three cofounders who is a hacker, so the coding would be solely his responsibility. He admits to "my own fear of getting mired into a huge development thing."

"What are the steps for you guys to launch another version of it for, like, rock?" asks Taggar.

"The big question for the start is, what domain? Can we buy Rock Genius.com?" The current owner of that domain name had not responded when the founders had e-mailed him.

"Trying to buy domains is a waste of time," says Taggar. "Don't go there. Trying to buy domains is one of these typical black holes that people fall into." Startup founders become convinced that they absolutely must secure a domain name, but as soon as the current owner of the name realizes how much a startup's founders want it, the owner puts a price on it that places it far out of reach.

"If we don't want to get RockGenius.com, we already have Rock Genius.net," says Lehman.

"This is a question that you have to figure out. At some point, you're going to have to expand into more stuff than just rap, right?"

"Right."

"So how does that look?"

"We could do Rock.RapGenius.com," Lehman says. "Or we could turn Rap Genius into a more umbrella name. The problem with that—

Mahbod is shaking his head—is you want to feel like the site you're on has personality. If you choose a totally generic name, some of the emotional appeal goes away."

Leaving aside the problem with the domain name, Taggar wants to know if they have given serious thought to expanding to rock lyrics.

"I'm itching to go," said Moghadam. "I'm the one who's always like, 'C'mon, we've got to do Rock Genius.'" He clowns, "I want to change my hair. I want to wear tighter pants."

Taggar suggests trying out a second genre in a modest way, by adding a rock section to the existing site. "I'm always a fan of, rather than trying to do huge, radical releases of stuff, just test out theories, the easy, simple way, and just see if anything starts happening," he says.

How should Rap Genius let users know that rock lyrics are available? Lehman suggests trying out a "Rock" tab on the front page, but he's not optimistic that a tab would solve the problem of clashing musical identities at a single Web site. "Ultimately, the name thing is stressing me out," he says. He has a point. With a RapGenius domain name, how far could it expand?

Moghadam revisits RockGenius.net. "Do you think .net—?"

"Don't do .net," says Taggar.

"What's so bad about .net, though? It's got sort of an indie feel to it, I thought, since indie rock will probably be central to Rock Genius."

"That kind of thing is exactly the thing not to overthink," says Taggar. "People understand .com."

Taggar has some advice for Rap Genius but presents it elliptically. "The way most people basically fuck up their three months of YC is by not doing stuff. If there's one way you're guaranteed to fail each week it's just not doing anything. That sounds ridiculous, right? We interview everyone, it's pretty selective. How possibly could someone get in and not do anything for three months? But we're not talking about people coming in and sitting around, drinking beer, and playing PlayStation for three months, right? It's people doing the wrong things. If you want to, you can spend all of your time refactoring your code or rewriting how you deal with database seeks or something—you could do that and that would be 'work,' in some sense of the word. But it wouldn't be work in the sense of getting you new users or anything that was actually important."

Taggar says that as he thinks about earlier examples of YC startups that tried a new product release or something else that had caused worry and anxiety before it was introduced, he cannot think of a single instance in which the outcome was "unfixably bad." He reminds them that experiments offer an opportunity to learn from users. "You try out something, people are, 'I liked *this*, but I didn't like *that*,' and now you have more information about what you should do next. Basically, inaction, not doing stuff, is the thing you should be worried about."

Taggar could not have been more diplomatic. He ends with an apology. "I appreciate that's kind of high-level, not very specific advice."

"Sure, sure," says Lehman.

"It would definitely be useful for you to speak with PG. Because the one thing that he's good at is the bigger-vision stuff—strategy—like how you guys could figure out a story that would be compelling to investors."

Taggar can push the Rap Genius founders in his gentle way to think about how to expand, Graham can push them in the same direction in his not so gentle way, all of the YC partners can say the same things over and over, but it is merely a suggestion, not a directive. In early summer, before they have taken on additional investors, the founders own 93 percent of their YC-backed startup and YC owns only 7 percent. If the founders do not wish to accept YC's advice, that is their prerogative.

8

ANGELS

To most everyone who lives outside of the startup world, the very name "venture capital" suggests that it operates at the far edge of risk. To startup insiders, however, venture capitalists are the most risk-averse investors in sight. In a 2007 essay titled "The Hacker's Guide to Investors," Graham paid tribute to the real risk takers, the individual angel investors without whose help most startups would never be able to build a business that would be of interest to venture capitalists. "A lot of people know Google raised money from Kleiner and Sequoia," he wrote, referring to the first and only venture capital that Google raised. "What most people don't realize is how late."[1] By then, the company could fetch a valuation of $75 million. Less well known are the angel investments that had come earlier, beginning with a $100,000 check that Andy Bechtolsheim, the first investor, handed Google's cofounders before they had incorporated and moved off the Stanford campus.[2]

Venture capitalists are "fast followers," in Graham's view. "Most of them don't try to predict what will win. They just try to notice quickly when something already is winning. But angels have to be able to predict." He had thought he would try his own hand at angel investing after he sold Viaweb, but seven years went by and he had not begun. Founding Y Combinator was his way of learning about angel investing. But he reduced his risk by investing in a batch of startups in one swoop—and as time went on, reduced it further by investing in larger batches. By 2009, YC had become too large for its partners to fund themselves, so it raised $1.75 million from Sequoia Capital in 2009, and a $6 million fund, most of it supplied by Sequoia, in 2010. Without planning to do so, YC had come to resemble a super-angel fund, making many small investments, as angels do, but deploying other people's money, as venture capitalists do.

Ron Conway is a well-known super angel who had almost three hundred active investments in late 2011.[3] He was an early investor in Google, Facebook, Twitter, and Zynga, and—through the Start Fund—all of YC's summer batch.

Another super angel is Dave McClure, whose 500 Startups fund, which raised $29 million in 2010, invested in more than 250 companies in less than two years. The accelerator it runs in downtown Mountain View is a much smaller scale than YC's, and most of its investments were unrelated to the accelerator. About twenty-five YC-funded companies were in 500 Startups' portfolio when McClure was asked at a conference what the principal differences were between his fund and YC's. He said Graham "would probably describe himself more as a home run hitter and I'd say we're more of a singles and doubles hitter," though he did hope for home runs, too.[4]

✦

At the very end of the summer, after Demo Day, when there is a last dinner gathering of the batch, Paul Graham will say something in passing that will give every one of the sixty-three companies pause. He says that at the beginning of the summer the Start Fund's backers, Yuri Milner and Ron Conway, had made $150,000 investments in startups that in almost every case were undeserving.

Graham delivers this news without a critical tone. He is being matter-of-fact, simply clearing up a mystery for the founders who do not understand at summer's end why other investors are balking at matching the same founder-friendly terms as they had received from the Start Fund.

"Let me explain," Graham says that evening. "It's *beee-caussse*"—he elongates the word to prepare his audience for the punch line—"Yuri and Ron's approach to investing in all of you, I'm sorry to say—sixty-two of you they invested in by accident. Statistically, there's an Airbnb or a Dropbox in here somewhere. And they don't know, especially at the very beginning of the batch, they don't know which one it is. So they've got to offer you all terms that sixty-two out of sixty-three of you don't deserve, to make sure that they get the Dropbox, whoever it is."

The Start Fund's investors knew that they had to offer terms that were literally irresistible. "If they had made their terms just a little bit harsh, in any way," Graham says, "there's a disproportionate chance that the *first* startup that would turn them down would be the best one. So they have to offer these ridiculously good terms for you."

✦

Graham and the summer founders often use the term "Start Fund" as shorthand for the two entities that, beginning with the winter 2011 batch, have invested separately in the startups. Strictly speaking, the Start Fund is now Yuri Milner's organization alone; it offered a $100,000 convertible note to every startup. Conway's organization is SV Angel, which offered a $50,000 note. The Start Fund and SV Angel are the only ones who make a no-questions-asked investment offer to every member of the batch and they enjoy high visibility in the life of YC. Yuri Milner gets an early slot in the speaker schedule and Ron Conway gets the next one.

The Start Fund and SV Angel are both interested in making additional investments in the most promising startups in the batch, what their partners will refer to as "doubling down." Milner employs a young associate, Felix Shpilman, to be his Silicon Valley representative. Shpilman, who graduated from the University of Southern California in 2007, happens to look younger than he is. He becomes a familiar presence in the YC hall on Tuesday afternoons before the dinners, darting from one clump of founders to another, setting up appointments to talk about possible follow-on investments.

Around YC, Ron Conway's name is used interchangeably with SV Angel, as in "Ron Conway's money," but Conway's formal position is "special adviser"—David Lee is the sole general partner, the one who actually manages it—and Conway was the contributor of the money only at the beginning. In 2009 and then again in 2010, the fund raised a total of $60 million from about one hundred wealthy individuals who are limited partners. For seed-stage investments, its $60 million will cover a lot of companies. With its investment of $50,000 in each company in YC's summer batch, it has created an index fund covering the entire batch at a cost of only about $3 million.

On the Tuesday that Conway will be speaking at YC after dinner, Lee, a former Google executive, and his junior associates come to YC earlier in the day to have short chats with the startups that are candidates for additional investment.

Lee sits down at the table where the four founders of MobileWorks are seated. The four, all of whom are coders, met while taking a graduate "social enterprise" class at Berkeley's School of Information. They have developed software for sending repetitive tasks to the cell phones of low-income workers in India. One of the company's guiding principles is to pay the workers at a rate that will help them climb out of poverty—crowdsourcing done in a way to make "crowd workers happy," says MobileWorks.

The MobileWorks cofounders are a multicontinental team. Anand Kulkarni and David Rolnitzky grew up in the United States. Prayag Narula came from India and Philipp Gutheim from Germany.

Kulkarni leads off, describing for Lee what MobileWorks does. "We act as intermediaries between companies and workers," he says.

"Got it," says Lee, as he writes notes in longhand in a bound notebook. "And how do you guys make money?"

"If you have documents you want to digitize—for example, you take notes in your notebook—that you'd like to have as a PDF, you upload it to our Web site, we'll chop it up into small pieces, send it to our workers, they'll digitize the words, send it back to us. We give you a PDF."

"Got it. How did you guys come up with the idea?"

Narula explains that they had originally thought they could develop software that would "scrape" the jobs listed on Amazon's Mechanical Turk service and then have the work done on the cell phones of workers in India. "But once we did our research, we understood it won't scale," he says.

"How do you guys all know each other?"

"The four of us were Berkeley graduate students," says Kulkarni. "It's spun out of research that we were doing as part of our graduate work."

"Are you live?"

"We're live with workers. Done about fifty thousand jobs so far," he says, referring to a pilot project with one corporate client. "We're not live on the Web." The software that clients would use for submitting the tasks they wanted done is being rebuilt. Kulkarni says the team hopes to have it ready in about a week.

"How's the division of labor?"

"All of us have a software background." Kulkarni says he does the pitching to investors; MobileWorks raised some money prior to acceptance at YC. The nominal roles of himself and the others are not particularly meaningful, however. "The reality is everybody's coding and building."

Kulkarni lists the sorts of tasks that MobileWorks will handle: "Handwriting digitization. Audio transcription. Language translation. Menial tasks. We're going to make easy tools that will let people integrate humans into the software."

"Can you educate me a little bit on the size of the opportunity—like how big is the Mechanical Turk business? I just don't know. Intuitively, it feels like a big opportunity."

Kulkarni runs through some figures, now estimating a market of $60 billion to $70 billion annually for outsourcing work that does not require special expertise and "can be done anywhere, can be done by anyone." One example is medical record transcription, a $15 billion market.

"So we're starting off in India," adds David Rolnitzky. "Primarily, it's because of the ubiquity of cell phones there. We talked about our social mission. There's a huge percentage of the population that's earning less than two dollars a day." Many do not have PCs but "they do have a mobile phone and an all-you-can-eat data plan. So it's very easy for us to reach out to people there."

Standard Chartered Bank, which has an office in San Francisco, is keenly interested in MobileWorks and has provided the startup with free office space.

Lee finishes his note taking, thanks the founders, then heads to his next appointment at a table on the other side of the room.

✦

If you ask David Lee if he thinks seed-stage financing is showing signs of a bubble, he vigorously disputes the suggestion. To him, encouraging everyone to start a business is healthy and harmless. Problems arise only if everyone is given significant amounts of money to actually build out unproven businesses. "*That*'s a bubble," he says. "That's like five million bucks for a consumer Internet company that hasn't even launched"—a reference to the way venture capitalists in the dot-com boom of the late 1990s threw millions at founders. He points out that the startups at YC do not have anything close to that kind of capital and do not hire in advance of demand. If the startups fail, the loss will be felt only by individuals who can afford to absorb it and who know all about the risks. SV Angel has no capital supplied by pension funds, university endowments, or other institutional investors. "What's the risk?" Lee asks. "The risk is a bunch of rich people"—the angel investors—"losing their money."

Lee has a master's in electrical engineering from Stanford but he also has a law degree from New York University. He practiced law and then joined Google, doing business development deals for Google's video initiative. Conway's career in tech started in the 1970s with marketing positions with National Semiconductor. Their SV Angel investments are primarily in startups that are "consumer-facing," but Lee favors technical founders.

"The biggest question for us, we look at a founder—I'm a forty-two-year-old guy—and I will say, 'Would I work for this guy?' " Lee explains how he evaluates founders when evaluating investment opportunities. "You just have to have that potential. I'm not saying you're going to have it from day one. I think people make the mistake of saying, 'This guy is no Jack Dorsey' "—the cofounder of Twitter and founder of Square—"or 'This guy is no Mark Zuckerberg.' " The people who actually met Dorsey or Zuckerberg in the early days before they became well known did not predict great things ahead for them—greatness is not obvious. "That's why we like the technical cofounder," says Lee. "'Cause he or she can build themselves out of a problem if their original pieces aren't working."

✦

It is time for David Lee's appointment with the MongoHQs, Jason McCay and Ben Wyrosdick. Lee slides onto a bench where they are seated and McCay begins with the story of how MongoHQ was born of frustration. There was no way to use MongoDB, the new database software, as a cloud service, running on someone else's hardware.

Wyrosdick picks up the story. "So we had to scratch our own itch there. We had to build something that we could use to provision these databases. We said this is something that a lot of people are going to need. Our goal at the beginning was to see if it was useful to other people. Turned out it was."

"It's always been a part-time gig for us," says McCay. "Now, when we did YC, that was the first opportunity that we've had to go full-time. We've been sprinting like crazy 'cause we never had the opportunity to do it full-time. We started about a year and a half ago, and we were the only ones out there."

"It's gotten very—" Lee begins.

"It's gotten very, very busy in a hurry, in the last two to three months. We're confident that we have the best product out there still in the market. We have ideas how we will continue to be the best. And we have a good head start. But now people are coming after us pretty hard."

"Who are some of those folks coming after you?"

"The biggest one, from a monetary standpoint, is MongoLab. Because they just secured $3 million in funding. And they've been in beta!" McCay is incredulous. "They're hiring people. They're getting ready to go. I *know* they're coming after us. They've done a lot—I say this as respectfully as I can to them, because I understand how it is—but they did a lot of basically taking what we did."

"So it sounds like you guys have built something for yourself. If I were to fast-forward five years, what is success for you guys? Let's say you're writing your LinkedIn or your résumé or whatever it is, you're saying what you've done in those last five years, what have you guys done?"

"I think for us it's like becoming a data layer for other applications. We're using MongoDB right now because we feel it's the—"

"—best of breed," Wyrosdick completes the thought.

Lee asks if they know Eliot Horowitz, who is the chief technology officer of 10gen, the company that is the developer of MongoDB software.

"He helped us a lot in the early days," says Wyrosdick. "Because before the stuff was even 1.0 we were trying to use it. And they were like, 'You can't use this in a shared environment.' 'Well, we really want to. So can you help us out?' So Eliot's staying up late, helping us getting patches into MongoDB, the platform, to help us be able to use it. 'Cause they really wanted us to be successful. Kind of brand awareness for them."

McCay returns to Lee's question about their long-term vision for the company and supplying the "data layer" to other developers. "Part of that strategy, at least near term, is trying to get into partnerships with all of these providers. We've been working with Heroku for over a year now. But also, there's all these Heroku-like startups now."

"What's the motivation in becoming the data layer?"

"I think it's because that's something that everyone needs. Every single application is going to need data. It's one of the things that most developers hate dealing with. Basically, you talk with developers—it's a pain to deal with their data. They want to be focusing on their code 'cause they feel like that's where they're innovating. Then what happens is they do something stupid and blow away the data or they don't have it backed up or something catastrophic happens and they don't have anyone watching that gate—and then they're in a bind. So we talked to people—even some of the YCs in this batch: 'Yeah, I have this thing out there but if I go and get TechCrunch'ed' "—meaning, get written up in TechCrunch and experience a spike in traffic—"'and this thing falls off the cliff, I'm screwed.' So we can go and say, 'Look, we have this provisioning platform. We'll monitor it, maintain it, keep it up. We're the experts in the field to help manage all that. All you have to do is worry about your code and make your product successful.' "

Wyrosdick tells Lee about selling pickaxes during the gold rush. Software developers "don't know how to make sure their data is secure and safe, all that kind of stuff. If we can sell the pickaxes and say, 'Go find the gold,' then it's the same model for us."

"OK, great," Lee says, signaling an end to the questions, but it's momentarily impossible to tell whether he likes the story that he has just heard. "Part of the reason I'm asking these little vague questions, and they're open-ended for a reason, is that I just want to get a feeling for guys as talented as you." Lee says, "I want to know why *this*? You know what I mean? Why spend the next five years of your life, or whatever it's going to be, focused 100 percent of your time on this? How did you get to this point? Because, as you probably know, it's easier than ever to start a business, but I think it's harder than ever to *build* a business."

Lee continues, "It's more important than ever to think about what you want to build. I like the vision of being that data layer. That sort of vision and aspiration is what gets the best engineers. You're going to spend a lot of your time trying to recruit people." If Lee's interest in investing in MongoHQ was not clear before, it is now. "So where are you in terms of fundraising? What's your process right now? Have you taken the Start money?"

The founders answer in unison: "Yeah."

"Awesome."

"For additional help on the back end," says McCay. "Like I said before, we were really trying to bootstrap the company. Our growth hasn't been explosive but it's been steady. We could show you the chart. We're hoping to do some stuff to shed some of our overall expenses." They explain that they are close to running at breakeven.

"As far as the funding goes," adds Wyrosdick, "we're taking their advice, holding off until Demo Day, spending these three months focusing hard on product."

McCay continues, "Part of our effort, too, is to rebuild this, but also we have a company to maintain. Keeping our servers going and everything stable and growing. As we start growing, things are changing and we have to stay on top of trends. Right now, I think we're 40 percent product and like 60 percent running the business."

"I'd like to invest on top of the 50K," says Lee. "To make your life easy, I'd just scratch out the 50K and write 150K. From what I've heard of what you guys are working on, I'd like to help you in any way I can. So think about it. I know time is of the essence. You guys don't want to screw around because Demo Day is an important milestone for you. To make it easy on you, just do it on the Start Fund terms. Which means if you guys do raise a round, the money is basically an advance payment on your round. But I just love the way you guys are talking about this. The sweet spot for us is developers building for themselves. That's what we like the best. So think about it."

Lee heads off to his next appointment. McCay and Wyrosdick confer and catch Lee before he leaves for the day. They happily accept.

A fifteen-minute chat has just brought in what they think will be an additional $100,000 and on the same terms that will leave other investors aghast when they hear about them. This is fund-raising at its easiest. Then it gets even better. The next evening, the MongoHQ founders learn that David Lee intended to give them not $150,000 in total but an additional $150,000 on top of the original $50,000. And he introduces them to another investor, who, after a phone call with them, will add $100,000. The one chat with SV Angel results in a $250,000 boost.

✦

David Lee will double down and make additional investments in about a third of the companies in YC's summer batch. But he regards his SV Angel's and the Start Fund's investment in every company in the batch as the more important piece. Asked if SV Angel's blanket investment was designed to secure the chance to get an early look at the startups, like MongoHQ, for those follow-on investments, Lee says no. Others have asked him about the underlying rationale for investing in every startup in the fund, implying by their question that it was devised to get publicity for SV Angel. Lee responds:

> I'm not doing this because it's a marketing ploy. I'm doing this because I think it's a great investment. Time will tell. It's an acknowledgment that it's really hard at an early stage to know who is going to be the Dropbox or Airbnb. If Dropbox doesn't scare the

shit out of every early investor, it should. It's not like they were a company everybody knew—"This is the cutest baby; we all need to get a part of this; let's invest in it." And it's going to turn out to be one of *the* defining companies in the next business cycle.

✦

The following Tuesday, Jason McCay and Ben Wyrosdick are at YC working at a table in the main hall before the group dinner. Paul Graham comes over, sits down, and asks a seemingly innocuous question: what have the two been working on?

"We've been doing mostly business stuff," McCay reports, apologetically.

Graham does not hear anything they should apologize for. When a startup is founded by an all-hacker team, Graham says he likes nothing more than hearing that the founders are working on the business side, "because you *know* they're doing it because they have to, because there's nothing they like better than to hack. The problem usually is getting them to work on business stuff."

The MongoHQs ask Graham's opinion of some venture capital firms that have reached out to them. He likes the firms that they mention. He says one will give the founders a quick answer. He praises another for being "among the non-golf-playing, younger VCs."

Graham does not say explicitly that he views MongoHQ as the top prospect of the summer batch, but by using the passive voice, that's the implication. "If you're considered the top one in the sixty-three, in this kind of funding environment, you can raise whatever you want." What would they do with a formal Series A round funded by venture capital firms, he wants to know. "Hire hackers?"

"Yes, hire hackers," says Wyrosdick.

"How many?"

"Three to four. And we want a business guy."

"Maybe you should raise 800K of angel money. Then at Demo Day say, 'We're not raising money. In a year we plan to raise a Series A at a very high valuation. If someone wants to give us the money now at that valuation, we'll take it.' So you can be perfectly relaxed at Demo Day. It'll be like you've already run the race and won the gold medal." He asks for a progress report: "What is your weekly revenue growth?"

Neither of the founders has the number. Wyrosdick begins tapping his laptop's keyboard to look it up.

"That's a bad sign that you don't know it," Graham scolds. "The gold standard of weekly revenue growth is 10 percent a week. That's insanely high. That works out to 142x a year."

He impresses upon them that they should know their growth rate and then obsess about meeting the target they set. "Treat it like a game. It will cause you to do the right thing. It will focus everything. It will be like a compass." The next time he sees them, he says, he wants to hear the number.

9

ALWAYS BE CLOSING

Chris Tam and Paul Chou are spending a lot of time in bars these first weeks. More than any other founders in the summer batch. Bars are where they must go to test out their ideas for Opez, a "fan site for bartenders," as they put it when they applied to YC. They would subsequently expand the idea to include other service occupations—waiters, hairstylists, personal trainers, masseuses, and others—in which customers sought out the best practitioners. Opez would offer a Web site that would allow "exceptional service professionals" to stay in touch with their customer-fans.

They had taken to heart Paul Graham's charge to all the founders to get out and meet their customers, and it was at a bar in Mountain View that they were pleased to see a bartender named Adam validate one of their hypotheses, that some service people are so good that customers will come to the establishment just to be served by them. Both Tam and Chou are hackers—Tam was a math major at Yale; Zhou majored in computer science at MIT. The two had met as summer interns at Goldman Sachs six years earlier and worked at Goldman three and four years, respectively. Tam has just earned an MBA at Harvard Business School, and both seem perfectly at ease introducing themselves to strangers and chatting animatedly.[1]

They had heard that Adam, the bartender, was a local legend and sought him out. After introducing themselves and describing what they were building, they secured his participation in a trial. The Opezes returned a few nights later with several friends and stepped back, dispatching the friends to the bar to ask for him. When he was asked for by name, Adam would put on a show. "It takes him eight to ten minutes to make these two drinks for our friends, which is kind of unheard of at a bar, right?" says Tam. The friends returned to the table and declared their drinks to be "fantastic." They pledged to return with more friends.

This was the memorable experience the Opezes wanted every customer to have. They also could see some potential problems. When there were four bartenders, for example, and a stream of customers came in asking for Adam and not for the others, there was resentment. The Opezes would need to think about how to promote the exceptional without antagonizing the not so exceptional.

The Opezes are working their way up and down the bars on Castro Street in Mountain View three and four evenings a week. Tonight, a Wednesday evening, the two are going to see what can be learned at a sushi bar, where they have set up their mobile office at a table. It's happy hour and again they are going to conduct market research in the field. They have printed up a business card for the occasion, which they have persuaded the manager to include each time the bill is presented to the customer. Across the top, the card asks, "How was my service?" and offers a free drink for submission of a review. The restaurant's name appears in one corner with the line ". . . cares about great service." And in the other, "Opez, finally great service."

The Web page to which they are directing customers looks like an iPhone app but it isn't an app—no download is required. Arriving at the Web site, the customer gets an Opez account and clicks to rate the server on a five-star scale and, electively, offer a comment. The restaurant's manager will receive an e-mail every time a customer submits feedback and will know if any server is receiving low ratings or negative comments.

By directing customers to a Web page, rather than have them download an app, the Opezes can make changes in the code on the fly while they sit in the sushi bar watching. They can see when a customer responds to the offer on the card and fiddles with a phone's Web browser; they can see the feedback as it comes in to their Web site. On their laptops, the Web site's code is displayed, ready to be tweaked—changes go live instantly. They tell the manager that they can immediately implement any idea she has, too. "We can build this tonight," they tell her. "Whatever you want."

Patrons receive their bills and the Opez offer. Only a few respond, however. Tam and Chou watch with sidelong glances as they type on their laptops. Tam, who is the more gregarious of the two founders, frequently hops up to confer with the manager. His girlfriend stops by; she came out from New York for the summer. Chou wonders whether the drink offer should be presented when a drink is served rather than waiting until the end when the bill arrives and the patrons are preparing to leave. The card could read: "If you want this drink free, why don't you think about doing a review?"

The founders know that offering the free drink to get customers is not sustainable. This is intended as a hack, a quick way to learn enough to be able to offer a service that is compelling on its own merits, without a sweetener. One thing they can see this evening is that waiters are not excited about the service the way bartenders have been. The servers see the feedback, which they know is being sent directly to the manager, as an instrument of management and a potential source of difficulty. And unlike a bartender, whose clientele is free to seek him or her out, a waiter does not, in most cases, draw similar loyalty. The server is not putting on a show.

The restaurant's patrons are not showing great enthusiasm for the opportunity to provide feedback, either. Tam and Chou have placed themselves amid prospective customers and see, instantly, when customers ignore their offer. It's rejection, what Graham and Livingston have warned the founders to expect. About nine out of ten patrons this evening are going to ignore Opez—right in front of the Opezes. Founders who sit in their apartment looking at traffic data for their forlorn Web site or download statistics for a mobile app that sits unnoticed in the App Store are relatively insulated from the sting of rejection. When you are sitting two feet away from prospective users when they are offered a financial incentive to give your startup a try and they still do not do so, the rejection is more personal. Tam and Chou seem unfazed, however.

✦

Many YC startups must engage in personal selling, especially those whose products are meant to be used by other businesses. In business-to-business startups, or B2Bs, the founders look for opportunities to chat up their startup wherever they happen to be. They must heed the imperative delivered in David Mamet's 1984 play *Glengarry Glen Ross*: "ABC. A. Always. B. Be. C. Closing. Always Be Closing. *Always!* Be Closing."[2] The circumstances in the case of these startups are different from those in Mamet's fictional setting, a dingy real estate office—the YC founders are self-employed and the atmosphere is as bright as Mamet's set was dark. But ABC is a necessity for all salespeople, including startup founders. Always Be Closing. *Always!* Be Closing.

Among the founders of the summer batch, Michael Litt of Vidyard is the one who most clearly was born a salesperson. He does not have to work on paying attention to sales. "Always Be Closing" comes naturally to him.

Vidyard, founded in Waterloo, Canada, offers to handle the behind-the-scenes details when companies place explanatory videos on their Web sites. Or it *will* handle the details, once it launches. In June, the software is still under construction, but the company is already one of the largest in the batch: including Litt, it has three founders and three employees. The plan is to offer businesses various options for videos, ranging from the least expensive, with a YouTube identifier that will be visible in the corner of the screen, to more expensive ones, without it. The Vidyard software, when complete, will track and report to Vidyard's clients how many people are watching each video each day; where they are located; whether viewers tend to stop watching it before reaching the end, and, if so, at what point; and other measurements. If the client does not have videos ready to go for placement on its site, Vidyard can produce them, made to order.

Litt sets up a conference call meeting with Say It Visually, a company in Bellingham, Washington, that produces explanatory videos for the Web sites of corporate clients. He hopes to persuade the company to send its clients to Vidyard to handle the video hosting.

Say It Visually cofounder Matthew Dunn asks Litt, "Talk a little bit geek to me about content distribution, storage redundancy, et cetera, et cetera."

"Absolutely," says Litt. "The back end was built on Rackspace and our content distribution was with Akamai. Recently we've moved over to Amazon S3 storage as well as CloudFront."

"OK," says Dunn.

"So our biggest thing was, we hated the fact that YouTube, Brightcove, the guys in this space—every once in a while, you click and it's just not loading. Or it would take five, six seconds. We actually saw, via the analytics we were pulling, that when a video is taking five to six seconds to load, 30 percent of the audience that clicked it was bailing before it even started playing. So we said, we're a bunch of engineering geeks. We want to optimize how fast that content streams." He pauses to laugh. "Our mentality was, if the content doesn't stream, or doesn't load, as fast as a Bugatti Veyron hits sixty miles per hour, then people are going to stop watching the video." (The Veyron is a sixteen-cylinder, thousand-horsepower automotive rocket that reaches sixty miles per hour from a standing stop in 2.3 seconds.)

The discussions with Say It Visually end without Vidyard landing new business, but its product offering is far from complete. It has built the infrastructure on the back end that serves videos reliably and fast. That portion of the work is complete. But it is still working on the video player that will go on the front end and on the software for the analytics. Litt has not waited for that work to be completed. Always Be Closing.

✦

Approaching strangers is not the first thing that hackers-cum-salespeople like Litt try. The founders of YC's B2B startups—which might be called D2D, developer to developer—first look for sales prospects among their summer batchmates. They are not looking for revenue, just beta users who can help iron out problems. With luck, these early users will become real customers in the future. Startups with database services—like MongoHQ—offer their services to fellow startups. So too does Envolve, a chat service that can be integrated into any other company's Web site, and

Paperlinks, which helps companies use QR codes in their marketing materials.

The founders of TightDB, Alexander Stigsen and Bjarne Christiansen, are not close to completing work on the TightDB software, but they have persuaded their summer batchmates, Michael Dwan and JP Ren of Snapjoy, the photo-organizing site, to try it. Snapjoy is itself a work in progress, and the startups' four founders work closely together, providing feedback that goes in both directions. As batchmates, they can get to know each other quickly, so Dwan and Ren trust that the TightDB founders will come immediately to their aid whenever problems with TightDB's unfinished software appear.

Summer batch startups could next expand their sales territory by knocking on the doors of YC startups from earlier batches. They know that the preceding YC founders had themselves been helped by still earlier YC startups and were glad to pay it forward, helping those just starting out, if they could.

If YC companies thought of themselves as an extended family, then Dropbox and Airbnb are, to the summer batch, the pair of wealthy great uncles, distant relations who might be approached for assistance but only once in a great while, delicately, and only after elaborate planning. Were either Dropbox or Airbnb willing to become a customer of a just born startup, it would be a boon of immeasurable value. But it is also difficult to even imagine the circumstances in which that would happen. Companies with wide name recognition would want to partner only with companies that had a proven record of reliable service and had earned the trust of other—sizable—business customers. Having a shared lineal connection to YC would not be a sufficient basis for the far larger company to rely on a not yet hatched startup.

When Michael Litt looks at Airbnb's Web site, which makes use of videos, he sees a natural home for Vidyard's services. Airbnb's AirTV section has videos of the more unusual lodging listings or colorful Airbnb hosts—its nickname is "the '*Cribs*' of Airbnb." Vidyard's software is not finished, but Litt goes after Airbnb, sending an e-mail to the cofounders, Brian Chesky and Joe Gebbia. It is the very first week of the summer session. Always Be Closing.

The subject heading of Litt's first e-mail message is "AirTV—Current YC company looking for some quick insight!" He opens, "Hey Guys, Sorry to bother. We're building a platform for video serving (encoding, analytics, customization included) and would love some insight into what you guys are doing with AirTV." He says Vidyard could add analytic reports to the videos that they are already serving or "we can serve and take the entire piece off of your shoulders." He closes with a mention that both Vidyard and Airbnb have shared parentage. "Thanks and let me know—we're very glad to be YC!"

His message is forwarded to Venetia Pristavec, the manager in charge of Airbnb's videos. Airbnb presently is using the video service of Vid Network, a startup that does the video hosting and analytics that Vidyard plans to offer. But Pristavec says that Airbnb is looking for a new video provider. Unfortunately for Vidyard, Airbnb has narrowed the field of replacement candidates to a pair of big names in video hosting, Ooyala and Brightcove. Clearly, Airbnb has decided to move away from relying on a startup and instead shift to a much larger company as its video supplier.

Litt is not daunted. He asks Pristavec if she has time for a chat. "It just so happens we're in a bid for another initiative with Ooyala and Brightcove," he says, suggesting implicitly that Vidyard is able to hold its own while battling with the big boys. He asks Pristavec only for information, for her "perspective" on the reasons for switching from Airbnb's current video host. "I promise I won't try to sell to you. I'm not a salesman—we're just looking for feedback from potential users," he says.

Meanwhile, Paul Graham dashes off a note to Pristavec and to Airbnb's cofounders, testifying to the Vidyard team's work ethic and asking that they be given an opportunity to make their pitch before a final decision is reached. Litt follows with an impassioned follow-up e-mail addressed directly to Gebbia. "This is an awesome opportunity that we are willing to work our asses off for," he writes. "It goes without saying that your brand would hold some amazing clout for us and the timing is perfect for both parties."

Litt dispatches more e-mails but it is all for naught. Airbnb does not budge from its original plan and Vidyard does not get the chance to make its pitch to land the contract. Litt does not pause. "Sorry to be a pain," he writes Airbnb again, confessing that "we're really putting on the hustle." He proposes that Vidyard produce some videos for Airbnb. This offer will not lead to a commission either. But to Litt, there is never a definitive no, there is only to-be-continued.

✦

Noah Ready-Campbell and Calvin Young, two software engineers who met at Google, arrive at YC with an idea that is so much bigger than a summer-sized project that it makes the ideas the other startups in the batch are working on seem like weekend projects. Their startup, Minno, is going to attempt to succeed where many other startups have failed: micropayments, the very small amounts of money that could potentially be charged for reading, watching, or listening to digital media. Other than building the software to handle the payments—the easy part—they must persuade Web site owners to ask their visitors to pay.

This has been a holy grail for publishers for as long as the Web has existed. In the 1990s, many startups tried, and failed, to get micropayments accepted: FirstVirtual, Cybercoin, Millicent, Digicash, Internet Dollar, Pay-2See, MicroMint, Cybercent. Clay Shirky wrote an epitaph in 2000, "The Case Against Micropayments," arguing that users hated them because they imposed a "mental transaction cost" as the user was forced to contemplate a transaction that would always be "too small to be worth the hassle."[3] Ready-Campbell and Young think the landscape has changed since then, that today users have become accustomed to paying small amounts of money for apps in the app stores and for virtual goods within games.

As an experiment, Noah Campbell-Ready and Calvin Young tried out their software on the *New York Times* Web site—without the *Times*' authorization. Users were permitted to hop over the *Times*' paywall by paying five cents to Minno in Minno's credits (users received two dollars in free credits when they registered at Minno). "NYT for a Nickel" did not last long; in just over an hour, they received a phone call from the *Times* instructing them to take the site down. It had served its intended purpose, drawing a bit of attention to Minno's revival of the micropayments idea.[4]

Investors are keenly interested in Minno. When the two founders arrive at YC, they have more than $700,000 in their pockets, raised from GRP Partners in Los Angeles and from angels. This is an amount that far exceeds what other members of the summer batch have raised at the start. It is so large that it dwarfs even the Start Fund's and SV Angel's $150,000.

Ready-Campbell and Young do not need the capital that YC offers, so why have they come? Their idea needs major Web publishers to adopt their payments system. Their $700,000 cannot purchase a meeting at Condé Nast. But with the YC affiliation, the Condé Nast meeting is set up quickly even before YC has its first dinner. This is accomplished with the help of Reddit, a member of the first YC batch in 2005 and part of the Condé Nast empire since being acquired in 2006. The meeting does not lead to a partnership, however.

Minno was the name the founders started with because the domain name Minnow.com could not be purchased for a reasonable price. But the two have decided to discard it, replacing it with BuySimple just as the summer session begins. They have also hired their first employee, Jeff Yolen, whose responsibility will be business development. In plain English, that means "gets a foot into the door of prospective content partners."

Yolen arranges a meeting in the first week of YC with mSpot, a company that has acted as wholesale supplier of radio programming to the major wireless carriers. It lately has added movies, too. It has also begun to build its own brand, renting movies at mSpot's own Web site that can be streamed either to phones or to a Web browser on a PC, competing with Amazon. Having built its core business as a supplier to persnickety carriers, it has to be larger than its unknown brand name might suggest.

MSpot seems to have a lot of executives. Yolen is old friends with one of them, Eric Thomas, who is an mSpot product manager. Yolen brings Ready-Campbell over to mSpot's modest offices in Palo Alto to meet with Thomas and two other managers, Brook Eaton and William Gaudreau.

"Jeff gave me the demo," Thomas tells Ready-Campbell, to get the meeting under way. "So I generally get what the product does. These guys"—he nods at Eaton and Gaudreau—"know at a very high level what it is but they haven't seen any of the details."

Ready-Campbell begins his pitch. "We want to make payments simpler for people to buy digital products on the Web. We think there's sort of been a big cultural shift in how people approach digital products. They do it a lot now. They do it in-app on the iPhone. They do it in social games on Facebook. And we want to build a payment system that lets any publisher directly monetize their audience." His delivery is smooth, as if he has done it a hundred times before, though he hasn't.

"We're in private beta now," he continues. "We just announced a partnership with Soundbug. They're a big indie music site, doing thirteen million uniques a month. And we're talking to a lot of others." He uses his laptop, which is connected to a projector, to show how the BuySimple payment service works for the first-time user. The screen shows a BuySimple button that sits next to a song title. "You just click here. You don't have to create a new account or anything. You just connect your Facebook account. Now, if I were logged into Facebook on this computer, I wouldn't even have to type—I would just see an 'Allow' dialog. Users don't have to verify their e-mail address or choose a new password or manage all of the issues of having a new account. Gets them on board really quickly. We pull in all the information about them from Facebook. We get their profile data and their social graph and things like that. That helps us manage our fraud a lot better. Assuming we think they're a real human, we give them two dollars just to start off with."

"What's the account behind it? How do they get money into their account?" asks Gaudreau.

"After you spend the two dollars or three dollars or whatever it is that you get when you first sign up, we prompt you for a credit card."

Gaudreau reviews: "So the concept is, the initial pay is hassle-free and then you go, OK, now come in and set up your account."

He moves on to another topic. "So I'm going to ask the question I'm sure you get every single time you talk to people: how are you different from PayPal, or better than PayPal?"

"Conversion rates is what it comes down to. The standard PayPal implementation involves opening up a new window, it involves entering all of your billing information, and then you have to come back to the original merchant site. There's three or four different steps. We think if you can remove three of those steps and make it one step, then it's a much better user experience and it becomes much more profitable for our partners."

MSpot's manager in charge of movie rentals, Brook Eaton, asks how much BuySimple would charge mSpot for handling the transaction. No charges at all, Ready-Campbell says. In the future, BuySimple would ask for a revenue share, perhaps taking 5 to 10 percent.

That won't be acceptable. "It needs to be better *and* save money," says Eaton. Right now, mSpot pays PayPal 5 percent, so BuySimple will have to come in with a much lower number to draw his interest. "For us, conversion rate doesn't even matter," he says. "'Cause at the end of the day, we need to make sure we run a profitable business."

As the mSpot executives look at one another, signaling the end of the meeting, Yolen looks at Eaton across the table and says with a straight face, "I know you're not interested in price." He's joking—price is the principal thing that Eaton is interested in. The two banter lightheartedly for a couple of minutes. When Yolen makes a tongue-in-cheek offer to lower BuySimple's price so low that BuySimple's would pay mSpot, Eaton turns serious: "I don't want you to pay me. I wouldn't want to do this if I didn't think you could be in business a long time."

"And get more users," adds Thomas. "This becomes a lot cooler when you have an account and use it for other things."

"Exactly," says Ready-Campbell.

As good-byes are exchanged, Yolen extracts a promise of hearing the following week whether mSpot is interested.

It will be a no, however. There will be many nos, and no yeses, from companies that matter. Salespeople need to be preternaturally optimistic, but they also need to pay attention to the no that signals that oblivion is ahead. BuySimple must contemplate the possibility that micropayments remain as premature in 2011 as in 2000.

During their first office hours with Paul Graham in mid-May, before the first YC dinner, Ready-Campbell and Young had been told that they seemed to be doing everything right. And Graham had remarked on the similarity not just in their clothes—both had on gray T-shirts and jeans—

but also in identical physical mannerisms, holding their arms the same way, something he had seen before among cofounders and had observed was a good sign, indicating greater likelihood of success. But he had also said that BuySimple's prospects were even more "binary" than most start-ups. Either it would succeed in a big way, or it would fall hard to earth.

10

CLONE MYSELF

CampusCred's founders had had a rough time during their finalist interview. They had what looked like a terrific idea—the deals site for college students. The problem was that they did not seem to Paul Graham to understand just how terrific it was.

Graham visibly became excited when the CampusCreds showed him an impressive graph of their startup's user growth in the six campus markets in California they had entered first. The rest of the country, however, was left untended, open for competitors to swoop in. This was a landgrab, and Graham could not sit still and keep his voice down as he talked. CampusCred needed to move quickly or it would lose out.

"You just colonize all of them," he had said, making it seem so simple to do. "There's no limit! Do them *all* next month!"

Now it was four weeks later, and they had added only two more campuses to their list. Thousands remained. The CampusCreds—Sagar Shah, twenty, the CEO; Brian Campbell, twenty-two, the sales guy; and Ben Pellow, twenty-eight, the hacker—would have liked nothing more than to work nonstop, forego sleep, and colonize in a month every campus in the country. They were already doing the working-nonstop/dispensing-with-sleep part before their YC interview, and they would continue to do so after. But so much work needed to be done. The CampusCred Web site needed to be rebuilt from scratch. The back-end code needed to be rebuilt, too. They needed a mobile app for students to use.

Pellow could not do all of this himself. Nor could Campbell personally call upon merchants by phone or carry out a land campaign, storefront by storefront, selling CampusCred deals in every market. He needed an army of clones as his field sales force.

CampusCred has an office in San Francisco. Once YC's summer batch has begun, the cofounders start coming down to Mountain View for office hours, seeking guidance on how to attack a seemingly endless to-do list. Shah has noticed that Paul Graham is best for the big, strategic questions, but Sam Altman is a better person to go to for questions about operational details. When office hours with Altman are posted for a Sunday afternoon, Shah is pleased to get the twenty-minute slot at the end of the day. He hopes Altman won't shoo them out when their appointment time is over.

✦

The first order of business is to bring Altman the latest news about how CampusCred is doing. It's near the end of the spring semester and the news is not particularly good. Shah hands Altman a graph showing sales. "The main problem that we're witnessing is that our penetration on a lot of these campuses—they're flatlining," he says. The average penetration rate—the percentage of the student body that has purchased at least one CampusCred deal—is about 8 percent at the campuses it has started selling at this semester; at Berkeley, however, where the startup began a year ago, the penetration rate is 15 percent.

"And what do you think is the max you could ever get?"

"Because of the student market, I think we could do like forty or fifty."

"That high?"

"Yeah, I really do think."

Campbell adds a qualification: "Depends on the school."

"And can you grow faster by working at existing schools and getting higher penetration rates or by launching new schools?"

"New schools," the cofounders say in unison.

"So is that your plan?"

"Yeah," says Shah. He brings out a map. "We're at eight schools," he says, pointing to their locations, all of which are in California, except for the University of Texas, Austin. Then he points to markings elsewhere on the map. "These are new universities that we've targeted that we want to go to, for sure, in the fall. These are high-priority targets. Making the total twenty." The number of schools that Shah wants to add—twelve—is more settled than the selection of which ones. He addresses Altman: "One of the things I wanted to ask you was, do you think it's better to go ahead and do kind of a national rollout in the fall? Or do you think it's better to do geographically located launches? For New York, for example, doing NYU and Columbia together. In Boston, Harvard, MIT, Boston University, and Northeastern all together?"

"How important was it to you when you launched your initial schools that they were in the same areas?"

"Not very," Shah and Pellow say simultaneously. Campbell is the one who had to sign up the merchants, however, and he does not answer.

Altman proceeds. "Then I would say you could go nationwide if you have the people to cover it. My instinct is that you're better off getting more schools at 7 or 8 percent than spending the effort trying to grow existing schools from 7 to 40 percent. 'Cause I think it's going to be a lot harder and it's also unproven. And you have proven you can get 7 or 8 percent" at newly added campuses. He continues: "It's sort of a landgrab, right? Like you guys want to win a lot of colleges very quickly."

Campbell now speaks up. "I think one of the biggest pieces holding us back is just the sales force. 'Cause launching, at least initially when you start out, it's all about deals that you have."

Altman asks, "Would you guys care if this document ended up on Andrew Mason's desk?" Mason is Groupon's chief executive, and Altman is holding up the multipage document that the CampusCreds have brought in today, listing the features that they want to incorporate into their redesigned Web site.

Campbell nods his head slightly in Shah's direction. "He would trip out."

"He would?"

Shah sheepishly agrees. He would trip out. "At least for now."

Altman resumes. "These guys are *so* distracted on their, you know, multi-hundred-million-dollar-a-year businesses, and they're going so quickly there, *and* they don't understand this at all—they are very unlikely to take you on seriously right now. Which is great, right? You could probably put this on the CEO of Living Social and Groupon's desk. That's unlikely to get the company to pay more than a passing glance at you. If anything, it ends up in an acquisition offer. Probably, it ends in nothing."

Brian Campbell wants suggestions about how to find the salespeople he wants to hire. "Our goal is to get maybe younger people that know the college market," he says.

"Yup! So the way I would look for this—the way most startups hire at this stage—is friends of friends. I would, like, all go home and post on Facebook: 'We're looking for someone in New York or Boston to do sales to businesses around college campuses. Do you know anyone?' One person can probably cover New York and Boston. One person can cover, like, mid-Atlantic. Does this work with a phone or do you have to actually walk into the stores?"

"It does work over the phone," Campbell begins. "But there are some businesses where it just won't work. If you want to get it done immediately, then you need to be in person."

What about other options? Shah asks Altman.

"Craigslist for this kind of thing actually works. The problem is you get a huge amount of junk. Have you guys posted on Hacker News?"

"No."

"That'd be my first choice. We had great results posting on Hacker News. You get really high-quality people."

Campbell is surprised. "Nontech?"

"Yeah. There are so many startup junkies that read Hacker News that just desperately want to be involved with a YC startup and aren't technical, and there aren't a lot of opportunities for them."

"What are some specific things we should look for or ask?" asks Campbell.

"You'll know the person when you see them. You're looking for someone who is like an animal, one of these people that is not going to take no for an answer. Super aggressive. Really scrappy. This is one of those things that's hard to define, but when you meet a good salesperson, you would buy something from them."

That's what CampusCred needs. Signing up merchants has become ever harder because "Groupon fatigue is really vast," Shah says.

A long discussion follows about the Web site's makeover and the expensive design and engineering contract that CampusCred is about to sign to outsource the work. Altman disapproves of the decision to outsource. "I'm itchy about it," he says, worrying about the risk that the code will be delivered late. He says he cannot think of another YC startup ever doing anything similar.

The chat has lasted more than an hour. Shah's hunch has proved correct: Altman speaks rapidly but is, at heart, a patient teacher.

✦

Brian Campbell posted a job notice on Hacker News and has received many responses. Some have come from young people the same age as Campbell, but many have come from older people, including some who have worked in banking or consulting. Campbell thinks they have a rosy view of startup life. It's "Google! IPO! Things are awesome in Silicon Valley!" he summarizes when recounting these conversations for his cofounders. Campbell would love to say out loud what he is thinking when he encounters this naïve view of startup life: "No, you've got to get your hands dirty and get shit done. To get to that level, there's a lot of hard work and bullshit lower-level stuff that you've got to get done. Because if you're not doing it, no one else is." Seeing older candidates' grievous ignorance of what work demands at a startup, Campbell is narrowing his search to recent college grads.

Today, he is on speakerphone with a candidate who is just finishing graduate school. Though he is not as old as the former McKinsey consultant that Campbell has also interviewed, he is not a fresh-faced twenty-two-year-old, either. Campbell wants to find "good talkers," but this candidate—let's give him a pseudonym of John Doe—does not have much to say. Nor does his voice convey animation.

"Hi, this is Brian from CampusCred. How are you?"

"Very well," says Doe.

"So, basically, got your résumé and wanted to get on the phone and chat with you a little bit. But have you heard about CampusCred and what we're doing?"

"Yeah. You're doing daily deals for everything campus oriented."

"Absolutely," says Campbell. He runs through a spiel, explaining how CampusCred began as a deals site, a Groupon for colleges, but is now looking at additional ways of connecting college students to retail businesses. He backs up a little bit. "That being said, deals are a huge part of what we're doing. It's the best way to enter into the market and capture it, if you will," he says.

Campbell talks rapidly in normal conversation, almost as fast as Sam Altman, but when he shifts into sell mode, he speaks doubly fast. The contrast between his rapid speaking style and that of this job candidate is extreme.

Doe says in a subdued voice, "I definitely understand the dynamic of how local businesses definitely try to engage with the ever-changing student population."

"Absolutely," says Campbell again automatically, a salesman's tic. "So I guess to cut right to the chase of what we're looking for, in the actual business development position, we're looking for people that can, basically, take ownership of a few regions. For example, the Northeast region—there's New York schools, there's Boston schools—we need someone to do basically what I've been able to do on the West Coast, which is calling up local businesses, get 'em excited about reaching students, going in the door, meeting them, getting them on board to try our platform, 'cause once they're on board, we have actually a really high retention rate with our businesses. 'Cause every year they want to continuously reach out to new students."

Doe interrupts. "What has your retention rate been thus far?"

"It depends a little bit on the campus itself," says Campbell. "I would say that out of every ten businesses, we usually only have one or two that really say, 'No, I don't ever want to do this again.' Then maybe one or two more that are like, 'Probably.' Which basically, to me, is, 'Yes'—I just have to get on the phone with them for an extra twenty minutes." He attempts to steer the conversation back to the candidate's suitability for the position. "Knowing the position and what it entails, I'm curious to see how you think you fit into that and experiences that back that up."

"I've been an active member of 'the market,'" says Doe. Other than being a student, however, he does not offer additional qualifications for the position. He wants to get back to the questions he has for Campbell. "How large is your merchant acquisition force?"

"So far, it's been me. So I've been able to go down and capture all these eight schools. And that's what we're looking to bring on"—he laughs in anticipation of what he's about to say—"essentially clone myself, and hopefully better. Make all the calls to those business owners that we *should* partner with, and want to partner with—if it takes meeting them in person, whatever it takes to get them on board with us and try it out and really become a long-term partner with us. So it's very, very aggressive, very competitive. We're looking for people that are self-motivated, if you will. Jumping back to you, I'm curious to know how you think you fit into that and if that's something you're interested in."

Doe does not speak up immediately. Campbell waits.

"That's interesting," Doe finally says. He has just finished graduate school and asks, "What kind of time commitment are you guys looking for?"

This is a strange question to pose, because, in startup life, commitment comes in only one size: total. "So we're looking for full-time and very much"—Campbell pauses to laugh a moment—"a lot of time." The laugh is a way of signaling that "a lot" is an understatement. "My guess is, especially starting out, when you're getting your feet wet, it's going to be pretty much nonstop. Obviously, it's a startup. So if it doesn't get done by yourself, then no one else is doing it. But the compensation itself—as a startup, we have to base a lot of things off of performance. We were looking at—we were hoping to get around $50,000 to $60,000 in terms of the salary, and I would say about $30,000 of that is base and the rest of that is based on performance and commission. I'd like to know your thoughts on that, where you fit in."

This has all come out in a breathless torrent. Doe pauses as he absorbs the news that the base salary is lower than he expected. The usual practice in a startup is to offer early employees significant slices of equity in the company, in the form of stock options, rather than market-level salaries. But Campbell has not mentioned anything about equity nor has Doe asked about it. Doe's attention remains fixed on the base salary.

"Um, I mean, it costs money to live and work in New York City," says Doe. "So that might be a little bit tough for me." Worried that he has shown less than exemplary enthusiasm, he attempts to muster a more positive response. "Nonetheless, I definitely understand the results performance. I think it's pretty cool—your forward momentum is definitely alluring." Doe says he is working on some other business projects at the moment, but would like to remain a candidate. "I don't know if I can fully commit all the time you need for this right now. But I would love the opportunity to keep talking as my schedule changes up."

"Absolutely," says Campbell. "Thanks!"

As he looks up, he sees Sagar Shah has come into the room while he was on the phone.

"I just walked in," Shah says. "His voice doesn't sound very sales-ready at all."

"No," Campbell agrees.

It was only one of many conversations like that he would have that day and following days. Prospective employees he spoke with had nothing close to the enthusiasm of the founders.

11

WHAT'S UP?

YC founders may know quite a few of their batchmates and what they are working on, or, more likely, they may know only some of them. It depends on how much time they have spent mingling with the others at the Tuesday dinners. Four weeks in, and the 160 founders are still randomly falling into "Hey, what are you guys working on?" exchanges with fellow founders who happen to be standing adjacent to them in the food line or drop onto the benches at the same table.

Shortly after the start of the batch, there was an official "party" put on by YC on a Saturday night, but it was not a party in any sense that an undergraduate would recognize. It was more of an open house, a sedate affair—organized around conversation, no music, early start, early end. It was the one chance for founders' romantic partners to visit and see the interior. Many founders elected to skip it.

Everyone has come today, however, the last Thursday of June. It's the first time that the founders have been directed to return to the YC hall just two days after a Tuesday dinner. The occasion is Prototype Day, when each startup will have two minutes to tell the group what idea it is working on. It's the day when everyone will systematically learn what everyone else is doing.

Before the mini-presentations begin, Paul Graham stands before the group and offers a short history of the Prototype Day tradition, laced with Grahamian exhortations. "Prototype Day is a trick," he begins. "It was invented by Robert Morris, who noticed the beneficial effects of a deadline on startups because of Demo Day. And he said, 'We should create this artificial deadline earlier on. As a way of spurring people to do things.'"

Today, however, this official rationale seems strange. These founders are not undergraduates who are prone to procrastination. They got into YC because they had a record of getting things done. They have been acutely aware, from day one, of the rapid approach of Demo Day—they visit a Demo Day page set up by a member of the batch that tells precisely the number of remaining days, hours, minutes, and seconds, followed by a scolding: "So why are you wasting time here?"

The real purpose of Prototype Day is to give all of the founders the opportunity to see what their peers are doing. "There's nobody here except you guys," Graham says. "Don't worry about looking good. Just explain what you're doing. And maybe some people will say, 'Oh, yeah, I had this friend who worked on that. This is what he did.'"

Two days before, at the dinner, Drew Houston, the founder of Dropbox—summer of 2007—had been the guest speaker. Afterward, he told Graham there was something else he had intended to say about successful startups but hadn't gotten to: "They don't fuck around, right? The startups that succeed, they don't go to meet-ups, they don't run around talking to boards of advisers, they just write code and talk to customers, right?" This is Graham's oft-repeated mantra, too. Write code and talk to customers.

Graham says the founders may not realize that he and the YC partners update their own rankings of the batch's startups as the session proceeds. It is a training exercise for the partners' own benefit, he explains. They want to improve their ability to identify the characteristics of applicants most closely associated with later success.

"I noticed," Graham says, "looking down the rankings—at least my rankings—that all the ones at the top were, like, real animals at doing sales." He does not mean to say that strength in sales was all that was needed. "You have to be good at hacking *and* be aggressive in sales. Everybody we fund is good at hacking. We can tell *that*!" What the YC partners might have missed is that willingness to attack sales. "When I say sales, I don't just mean people calling you up. I mean going out and talking to customers to figure out what they want. So why don't you try this: erring on the side of sales. Just spend *all* your time doing sales and treating hacking as this side project, right?" The outrageousness of his suggestion makes Graham more animated. "*That*—" He pauses for emphasis. "*That* will tend to produce really good results."

Graham has an example in mind. He looks out, searching. "Vidyard—where's Vidyard? Vidyard tried to sell its services to Airbnb, and Airbnb said no, they had already signed this contract to write code or something like that. Vidyard said, 'OK, we'll nag Airbnb anyway 'cause I know something is going to happen some way.'" His exhortations reach their climax: "Be more like Vidyard. Be sales animals. And if you're not sales animals, force yourselves to do it, even though it will be uncomfortable."

✦

A laptop is set up on a table whose screen is projected on the white wall behind the dais. Though today's event is called Prototype Day, the prototype need not be shown. The batch is too large for each startup to introduce its idea and still have time to run through a full demonstration. A few well-presented words about the idea should be sufficient.

For some teams, their idea remains rather hazy and the startup seems a long way from getting started. Or, in the case of others, the startup already has real customers and has grown well beyond a prototype stage. Most of the startups fall somewhere in between the two extremes.

The Kalvins, the three young men who had applied with the idea for printed photo books and about whom Paul Graham had privately noted, "Insanely energetic founders. Fund for the new idea," have found the new idea. They call their startup Ridejoy, after trying out Ridebank, Ridetastic, and Ridebee. Jason Shen introduces the idea. "We're everything for rides. We want to change the way people travel by making it super easy and fun to share rides with other people." The number of rideshare posts on Craigslist—about a thousand are listed at the moment for just the San

Francisco Bay Area—provides encouraging evidence of demand. Shen says they will start off with long-distance ridesharing, to the upcoming Burning Man festival in the desert and also between San Francisco and Los Angeles.

As the other startups do at the end of their presentations, Shen offers to the batch the expertise of his team's members: "Kalvin and Randy are developers," he says, and as for himself, he knows how to stay motivated in the face of rejection. "I've gotten rejected thirty days in a row," he says, a reference to his putting himself through "Rejection Therapy," in which one must make unreasonable requests so that one is rejected by a different person, at least once, every single day—inuring one to the pain of rejection.[1] (One example of Shen's bid to be rejected: he asked a flight attendant if he could move up to first class for free. In another case, he saw an attractive woman on the train and decided he would ask her for her phone number, and when she would turn him down, he would have fulfilled the day's required quota of rejection. He sat near her, fell into a conversation, and when they got off the train and he asked for her number, she said, "Sure." He categorized this as "Failed Rejection.") "So if you need to get pumped up for your sales calls, talk to me."

The BrandonBs—Brandon Ballinger and Jason Tan—also have a new idea. They listened when the YC partners discouraged them from trying to do a consumer product. Ballinger says, "We're Sift Science and we're the sheriffs of the Internet." For a touch of levity, behind him is a projected picture of the cast of Comedy Central's *Reno 911*. The founders are working on software that sifts through "suspicious user activity," looking, for example, for spammy comments on online discussion forums, which had been their original idea. Or the software might look for outright fraud at a site like Airbnb, where a fraudster, registering as both the host and guest, could use the site to launder money. Ballinger closes with a pitch to his batchmates: "If you have any of those problems, come talk to us. We'd love to help you."

After another founder, who is from India, ends his presentation, Graham addresses the group from the side of the room. "In the course of these talks, things will come up that might be useful to remember for Demo Day. And the lesson for that in this one is: if you have a foreign accent, you have to talk slower."

One of the larger teams is Parse, with four cofounders. They include two experienced YC founders: Tikhon Bernstam, who cofounded Scribd, in YC's summer 2006 batch, and Kevin Lacker, who was a cofounder of Gamador in winter 2010. Bernstam, speaking as a developer who will be selling Parse's service to other developers, opens with a crisp description: "We're Heroku for mobile apps." This is the high concept invariably invoked by YC founders when their startup ideas concern cloud services.

The Parse team, a merger suggested by Graham of two teams that had applied separately, had not gotten started before June but has been moving quickly since. "Our site is live already. Just by dropping in our SDK"—the software development kit—"you don't have to write any back-end code, you don't have to wrangle with Rails or PHP. We have all the cloud services that you'd want for your mobile app." In just three weeks, the startup has written software development kits for both Apple's iOS devices and for Android, and launched a private beta serving two hundred developers, some of whom have already publicly released apps that rely on Parse. Bernstam returns to his pitch. "It's really this easy: drop in the SDK, and with three lines of code—literally three lines of code—you can be storing data back to the Parse cloud." He lists investors that have put in money since the start of the summer, including Google Ventures. This is the fastest start of the batch.

Graham calls out to the other founders: "Did you guys notice how easy his presentation was to understand? As soon as he talks slow enough—just notice, almost everyone talks too fast—if anyone talks at the right speed, it's so much more easy to understand. *Talk. Slow.*"

Adpop Media's software technology, which digitally inserts product images into video, demands to be demonstrated, with Yin Yin Wu presenting. She shows a video of a tennis match with a giant "NIKE" apparently painted on the court. It looks like it is actually affixed to the surface of the court, but it's an illusion, created by Adpop's software after the tennis match. The software runs on ordinary computer hardware, not the specialized equipment needed to draw a football telecast's yellow line of scrimmage across the field. The software is easy to use, she says. "Select four points in one frame—the area where you want an ad or image—and we pump out another video that has the ad inserted in it. It looks like it's part of the original."

✦

One criticism of Y Combinator is that the ideas that founders pursue tend to be lacking in ambition, that startups aiming for a Demo Day in one hundred days tend to think small, that Dropbox's Drew Houston, who picked a big market, is an anomaly. As the presentations proceed this afternoon, there are a number of startups that do present smallish ideas, prompting the question: how would these founders' ideas ever grow into a substantial business? One startup that seems to fall into this category of exceedingly modest ambitions is Splitterbug. The two founders, Matt Holden and Sean Lynch, are former Google program managers, so they know something about companies that think big. But Splitterbug is a mobile app that does nothing more than handle the sharing of expenses, for example between roommates or between friends who are taking a trip together.[2] No one in the audience criticizes the idea, however. When the Splitterbugs finish their presentation, they receive the same polite applause as every team that has come before them.

When selecting startups to fund, YC's partners pay more attention to founders than to their ideas. This means some startups in the batch are likely to have overlapping or nearly identical ideas. This summer, no fewer than three startups begin with ideas for travel apps and two have ideas that involve serving up technical challenges to programmers. This is not a matter of great concern to Graham and his partners. They leave it to the startups to sort out. If one startup feels crowded by another, its founders are free to take on a new idea.

Interview Street is one of the two startups working on testing hackers. It has settled on screening job candidates on behalf of software companies, which will pay Interview Street when they make technical hires. Its Web site is already up and running. Another startup in the same batch, which does not yet have a name, also plans to offer programmers technical challenges. But it would not screen on behalf of particular employers. Instead, it would give programmers the chance to establish their technical reputations by solving gamelike challenges.

Ryan Bubinski and Zach Sims are the cofounders of the second company. Both were Columbia students. Bubinski, a double major in computer science and biophysics, has just graduated. Sims, a political science major, has just finished his junior year. Bubinski had founded a club, the Application Development Initiative, which arranged for students to teach programming to fellow Columbia students. Sims, who had taken only one computer science class, was exactly the sort of nontechnical student that the initiative sought to help. The club also offered to give computer science students more practical skills than those they were taught. Academe has always been slow to change the programming languages it uses in the classroom. Today, computer science courses still heavily favor Java, largely ignoring the languages most commonly used in Web development, such as JavaScript and either Ruby or Python. Bubinski's club offered to remedy this.

Sims is an anomaly among the founders in the summer batch: he is a twenty-one-year-old nonhacker who already has more than a year of startup-related experience under his belt. A year previously, he had had a summer job at AOL Ventures, the venture capital arm of AOL. Then, during the previous school year, he had talked his way into being the first hire at a startup in New York called GroupMe, where he worked "part-time"—thirty to forty hours a week. (His choice of employer turned out to be most fortunate, as it gave him an insider's view of startup growth at its fastest. GroupMe, which offered texting to groups, launched in August 2010, funded by an $850,000 seed round; in January 2011, it raised $10 million from venture capitalists; and before summer 2011 is over, it will be acquired by Skype for more than $80 million, just one year later.)

When Sims and Bubinski arrived at YC and met other members of the summer batch, they discovered Interview Street, whose idea was not exactly identical to theirs, but close. The most dispiriting discovery, however, was what hackers in the batch told them about their plans to gamify programming challenges:

I would never ever use this.
Why would a good programmer try to prove their abilities this way?
What good programmer is actually struggling to find a job?

Bubinski and Sims decided to look for another idea. They were renting an apartment in Sunnyvale, a place far duller than Mountain View and even farther from San Francisco, and discovered when they attempted to order food that many restaurants did not have their own Web sites. Today, the two introduce their new idea: BizPress, a service to give small businesses an easy way to set up their own Web sites.

"We did a quick survey on Castro Street in Mountain View," says Sims. "There are twelve businesses on a block, and out of twelve, six of them don't have Web sites." BizPress would fill this apparent need, offering a free home page for small businesses now and optional fee-based services, like accounting and social media management, later.

BuySimple, with its micropayments system, is the startup with arguably the most ambitious idea in the entire batch. The startup does not present today, however, because BuySimple is no longer part of the batch. Its founders have come to the reluctant realization that the online world is not yet ready to accept micropayments for content, so the two have withdrawn from YC.[3] The total number of companies in the batch has dropped from sixty-four to sixty-three.

The batch is not devoid of big ideas, however. Clerky stands out among the presentations as one of the more ambitious. Darby Wong, one of its two cofounders, is talking about its target market and has an attention-getting figure to go with it: $25 billion. Wong and his cofounder, Chris Field, are attorneys who are seeking to automate routine legal transactions in business. Handily, they're not only attorneys—they're also software engineers. "We think of ourselves as Heroku for business transactions," says Wong. The two Clerkys are building the software infrastructure that will be able to handle any legal transaction entirely online. Where LegalZoom has predefined forms for only the most basic transactions, Clerky will let lawyers and businesses upload their own forms for any kind of transaction. And where LegalZoom prints and mails back paper documents for signing, Clerky will use electronic signatures, making "paperwork" an anachronism. Wong lists examples: employment agreements, nondisclosure agreements, stock option agreements, supplier agreements used in manufacturing, lease and rental applications handled by landlords. "This is where Clerky comes in. We streamline and automate those transactions to make things as easy as buying something off Amazon." This is not airy talk. Every startup in the room has already used Clerky: the module for convertible notes was ready for real use at the beginning of the summer, when the Start Fund and SV Angel used it to handle the convertible notes.

✦

Some of the presenters are difficult to hear. Some have little to show at this point. But the Rap Geniuses form a category all their own, founders who revel in the opportunity to ascend a stage. Tom Lehman, Ilan Zechory, and Mahbod Moghadam assemble at the front of the hall for their turn.

"What's up, guys? I'm Tom. This is Ilan—" He motions to Zechory, who is sitting at the laptop connected to the projector.

"What's up?" Zechory echoes.

Lehman continues, "And Mahbod, and we are Rap Genius. We are building a lyrics site that doesn't suck." He throws out a line from the Lil Wayne song "Six Foot Seven Foot" containing a bizarre reference to lasagna. "What does that mean?" he asks. No one answers. "So type into Google and see, OK, what site has it?" Zechory has typed the line into a Google search box and a page of results are now showing. "We're the first result. You click it. And it takes you to Rap Genius, takes you to the song page and will actually open the explanation for the line right here. So you can read a little bit about it." Lehman gives a quick tour of the Web site. Then a rival's Web site is shown: MetroLyrics, which is affiliated with AOL. "This is our competition," Lehman says in a contemptuous voice. "This is what a lyrics site looks like right now. You go, you see a big ringtone ad. This is terrible! Where can you find the lyrics?" He mocks the ads that fill the site; the audience laughs at the absurdity of an ad for breakfast cereal that has inexplicably been placed by Lil Wayne's lyrics.

Lehman's voice grows louder and confident. "So we're going to basically dominate the lyrics page. We're going to kill MetroLyrics, kill all these sites. After that, just all texts. Poetry. The Bible. Literature. Tax code." The audience laughs again. "You got it." The other areas are mentioned in a blur, with no mention of what will actually be next after rap lyrics. Strangely, there is no mention of work on rock lyrics, even though it has been six weeks since the founders had the office hours with Harj Taggar in which they seemed ready to try that as their second area.

✦

The most playful presentation of the day is also one of the best received. Launchpad Toys' two cofounders, Thushan Amarasiriwardena and Andy Russell, have released an educational game for the iPad called Toontastic, which young children can use to create an animated story. It would not immediately seem to be a product of interest to this room of hackers, most of whom do not have children. But the two founders have a business story to tell, in addition to demonstrating the Launchpad Toys software. With Russell sitting at the laptop, Amarasiriwardena begins with an image of a figure built with Legos. "There's a real interesting thing going on in the toy industry right now," he says. "The top three toys that kids are asking for aren't really what we'd consider 'toys.' Duracell puts out a report every year of what kids are requesting and the top three are: an iPhone, an iPod Touch, and an iPad." The audience laughs, surprised at what they see. "That gives us a humongous opportunity," Amarasiriwardena explains. "There are two markets hurtling toward each other: that's the creative play industry, which is like Play-Doh, Legos—which is $3.5 billion—and the kids' video game industry, which is $2.5 billion." Combining creative play and children's video games produces a new category, which he calls "digital play."

"What does digital play look like? It looks like our app Toontastic, which has been out in the App Store for a couple months. We've been featured a number of times. One hundred thousand cartoons have been made on the device." He holds an iPad that is now connected to the projector and he demonstrates as he speaks. "It's really simple. It's action figures—you can scale, you can rotate, move individual arms, they walk in different directions." This is a break from the other presentation fare, of developers selling to developers, and the crowd laughs at everything the Toontastic characters do.

Amarasiriwardena explains, "What we're trying to do is take the play patterns that every kid is totally familiar with and just digitize them. Just add a record button. This summer we're going to work on building virality. That's one of the key ways of growth for us, by sharing cartoons. Every parent loves to show off what their kids are doing. We're adding much stronger social tools into our app." Launchpad Toys will also start offering virtual goods for sale that can be used within Toontastic. "Digital items cost nothing to make, really," he says.

Amarasiriwardena ends with a quick demonstration of the second app that he and Russell are working on, which children can use to make their own videos. It combines the functionality of the video camera built into the iPad with an animated overlay. It does not yet have a name, but he demonstrates, pointing the iPad's camera at the audience. A drawing of a ray gun is seen in the foreground and a live picture of the crowd is in the background, sitting targets for the gun. "I am going to blast you," he says in a bass stage voice. The software emits electronic noises for the blasts. He returns to his regular voice and winds up, "So we're going to have tons of little toys like this." The audience applauds loudly. It likes Launchpad Toys.

When the applause has died down, Paul Graham, standing at the back, shouts out, "By the way, *that's* a really good presentation."

✦

After three hours of presentations, it is time for the group to vote. Each founder is to send in two text messages to indicate the two favorite startups—excluding their own, of course. A few minutes later, Graham announces, "We have winners." He begins with eighth place and works his way up.

"There's a tie for second between Parse and Adpop, with twenty-two votes each. And number one—this is really close—with twenty-three votes: Launchpad Toys!" After the applause subsides, he ends by imparting a brief exhortation. "So what do you remember about these people? A good presentation gets people's attention. Graphs that go up and a good presentation."

12

HACKATHON

To an onlooker, the work of software engineers is utterly lacking in visual interest. Depending on their nocturnal nature, YC's founders tap at keyboards all day or all night long. Had YC adopted an incubator model and had a single, hangar-sized work space for everyone, the sight of 160 founders in the summer batch working shoulder to shoulder in a bull pen still would not have offered much to hold a visitor's interest for long. Nor does anything like such a scene exist. By design, YC's founders work at their own places, usually living quarters, scattered around Silicon Valley, just like grad students. The founders' experiences are their own.

For Snapjoy, the long-awaited day of its public debut has arrived. TechCrunch has run an article introducing the service, new users are signing up for the free "public beta" offering, and the stability of its software is undergoing a severe stress test.[1] Founders Michael Dwan and JP Ren, the hackers from Boulder, are anxious. They have stayed up until four thirty this morning debugging code. They fret that their site will be overwhelmed by a surge in traffic and they will have another long night ahead. They are prepared: they have stocked up on Red Bull.

Their apartment is in Sunnyvale, about a ten-minute drive from YC, in the direction away from San Francisco. It's a location suited for round-the-clock work, without diversions. The apartment is minimally furnished and without a television set, a place lived in by sojourners. All of the walls in the front room are covered with giant sheets of paper that serve as whiteboards, outlining Snapjoy's software architecture and to-do lists. Dwan says that someone looking in from the outside at night must think "we look like terrorists 'cause all we have on the wall are these plots for world domination."

One sheet lists milestones in the cumulative number of photos that users have uploaded to the site and the time at which each milestone is reached. Looking at his laptop, Dwan gives a whoop and hops up to add a new entry on the sheet: 100,000. The achievement is memorialized by a camera that sits on a tripod; it is taking a snapshot of the living room every ten seconds for eight hours, creating the materials for a time-lapse video that they will prepare later. Dwan's wife oversees the camera. Their cat walks along the back of the couch, not deigning to acknowledge the achievement.

Dwan and Ren say they experienced two "freak-outs" earlier that day when problems surfaced. The most recent one was caused by a corrupt file that a user had uploaded. This caused havoc amid the processing that was being done for Snapjoy by Heroku, and thirty-six thousand uploads quickly backed up, waiting to be processed. Snapjoy needed to increase the number of "dynos," or units of computation, that Heroku assigned to it. They filed a support ticket, but the allocation of one hundred remained unchanged. Dwan sent an instant message to a Heroku founder, "Please help a YC brother out!" and quickly got another hundred dynos. The queue fell back to zero.

The TechCrunch story rides to the top of Hacker News, which brings in more users. At one point, they see one thousand photos coming in a minute. They are going to quickly burn through the $50,000 in credit that Heroku gives every YC startup.[2] (YC's startups also receive credits from other providers besides Heroku, including Amazon Web Services, Dropbox, Rackspace, Mixpanel, and Microsoft.) Ren says they're a little worried: "Heroku e-mailed us and said, 'You guys have a lot of credit from us. You're probably fine for a few days.'" A few days? Then what?

Dwan and Ren expect that the users who come to Snapjoy are going to upload personal collections of five thousand photos each. When each one arrives after uploading, Snapjoy's software looks for the identical photo in a higher resolution uploaded by another user, such as a family member. If one is found, then the higher-resolution version takes its place in the collection of everyone who has a lower-resolution one.

The software is holding up. Ren looks at his laptop, which displays error messages. Nothing critical. "It seems that the errors that we see are just a by-product of two 'workers'" ("workers" is an anthropomorphic term used by Heroku to refer to its software; so too is "swarming the job"). "The system is built so we never really lose data. We'd rather reprocess something twice than miss something." They have not yet had time to write the code to detect a corrupt file before it is uploaded.

As long as Snapjoy labels its service as "beta," it will be offered free. Prospective users are told that when Snapjoy begins to charge for its service, it will probably cost about one dollar a month for four gigabytes of photo files. The Snapjoys are pleased to see from the software that monitors what visitors are doing on their site that users are signing up even after visiting the page that explains the plans to charge in the future. Ren thinks that users who are considering entrusting their personal collection of photos to a service actually want to pay for it—how trustworthy would a service appear if it had no visible means of sustaining its service?

The two may joke about their plans for world domination, but they are earnest when they talk about longer-term goals: getting a million users, building a pair of data centers of their own so that they don't have to use Amazon's relatively expensive cloud storage service, S3. Once they have collected everyone's photos, they want to add everyone's videos—each one might require a gigabyte of storage. An individual would need many terabytes (a terabyte is one thousand gigabytes of storage) or even petabytes (each is one thousand terabytes). Their plan envisages backups of backups of backups, multiple redundancy that entails nearly bottomless storage needs.

At the moment, though, uploads of users' photographs are limited by the bandwidth of their Internet connections; it will take days for an individual to move the family's collection of photos up to Snapjoy. At their own apartment, however, the Snapjoys have an extremely fast Internet connection. "It's ridiculous," Dwan says. "We can upload a three-meg photo every two seconds. A seventy-megabit connection. Comcast business class, which we got through YC." He goes "beep, beep, beep," simulating the speed of photos zipping through. "We're really going to miss this when we leave here." An out-of-town friend who stopped by their apartment for a visit and saw for himself how fast the connection was told them after he returned home, "You guys are not living in reality." Dwan concedes the point. "It's actually difficult to perceive the site as normal users do," he says. "So we need to go to a coffee shop. 'Cause what looks really fast and smooth to us is not the same for most people."

✦

When Paul Graham and Robert Morris started Viaweb, they used Morris's apartment and Graham slept in the bedroom of an absent roommate. Graham's advice to founders is that they should do the same: use their living space as their office. "Ever notice how much easier it is to hack at home than at work?" he wrote in his 2005 essay "How to Start a Startup."[3] From the beginning of YC, he and his partners have not provided work space. This is not to minimize expenses. "We did it because we want their software to be good," he explained. He mocked the "professional" notion that work and life are supposed to occupy separate spheres.[4] In a startup that begins in an apartment, the founders

> work odd hours, wearing the most casual of clothing. They look at whatever they want online without worrying whether it's "work safe." The cheery, bland language of the office is replaced by wicked humor. And you know what? The company at this stage is probably the most productive it's ever going to be.[5]

Almost all the founders in YC's summer batch, like the Snapjoys, work at their apartments. There are times, though, when founders crave having more company around than that supplied by one or two cofounders. As the summer batch progresses, the startups based in San Francisco organize an occasional hackathon, a designated evening in which others in the batch are invited to come over with their laptops to the host's space to work, share solutions to common problems, and trade tidbits of YC-related news.

The smallish apartment that most companies occupy would not be suitable, so it is fortunate that a few YC startups in the batch have deviated from the recommended path and are occupying office space. MobileWorks, which has space in the ground floor of Standard Chartered Bank, hosted a hackathon a week ago. Tonight, it is the CampusCreds who are hosting, and they have many thousands of square feet to share with their batchmates. Their "office" consists of a few tables among many in a large open room shared with many other startups. All are renting space on a "per desk" basis from a company called RocketSpace.[6]

✦

This evening, CampusCred and its guests have the top floor to themselves. The first person to arrive is James Tamplin of Envolve, which provides a chat service used on the Web sites of other startups. A few minutes later, three MobileWorks cofounders—Prayag Narula, Philipp Gutheim, and David Rolnitzky—come up, followed by Sift Science's Brandon Ballinger and Jason Tan. Others would come later.

Tamplin opens his laptop and pulls out a little bottle of 5-Hour Energy drink, which he explains is keeping him going after getting only three hours of sleep the night before. He does not expect to get any sleep at all this night either, as he and his cofounder are preparing to push out a new version of Envolve's software the next day.

Seeing the energy drink, Ben Pellow, CampusCred's hacker-founder, raises a bottle to show everyone what he relies upon: B vitamins. There are occasions, though, when vitamins aren't enough. He pulls out a sizable glass jar whose interior is murky. "How about this? This is the hard-core shit," he says, laughing. "That is Siberian ginseng. These little capsules are filled with oil and ginseng. An energy thing." He's not done showing off his armamentarium. "If neither of these work, there's the third, super go-to, fuckin' insane"—he is now holding another jar—"but I don't like taking it 'cause I get *so* wired."

He's asked if it's legal, and he says it is. It's just a tea. He reads from the label: "'Increasing energy, stamina'—blah, blah, blah—'blood circulation, alertness, anti-hypersensitive, anti-diabetic, anti-ulcer.' That shit is almost impossible to even choke down. But it's super powerful." He laughs at the thought of how well it works. "So, yeah, that's how you do eighty-, ninety-hour weeks for a year. You don't ever get tired." He laughs again. "And I'm still alive."

Tamplin looks around the empty floor, where tables are filled with computer monitors for CampusCred's neighbors, but no one is present. "It's only seven o'clock on a Thursday. Where the hell are all these other companies? Slackers!"

Prayag Narula sees Pellow's Berkeley T-shirt and is surprised. "I didn't know you were from Cal."

"I am!"

Narula, peering at his laptop screen, reads aloud the headline of a story on Hacker News. "'MongoHQ Acquires MongoMachine.' Wow!" He laughs.

Jason Tan says, "I saw that. I think it was a one-person acquisition."

"We are four people," says Narula, referring to MobileWorks' co-founders, "and we have so much trouble making decisions."

"That's why we're going to have one consciousness: PG," says Tan. He adopts a deep monotone, pretending to be Paul Graham acting as an oracle: "I envision the future of advertising is—Go, my minions!"

Philipp Gutheim returns to Narula's point about four cofounders being unwieldy. The reason he says that he and his cofounders use MobileWorks as their startup's name is not because they're especially fond of it but because it would be too difficult to marshal a unanimous vote for an alternative domain name. He recalls, "We were lucky to be able to buy 'MobileWorks'—and now we're launched—so end of discussion." He laughs.

Ballinger, who previously worked at Google, says, "That's how big companies work, too. Nobody wants to piss anybody off with changes."

Tan, who has set up his laptop at the far end of a long table, wanders down to the other end, where the CampusCred founders and sundry student-employees and student-consultants sit.

"You guys should acquire MongoHQ. Six persons take over two persons," he jokes.

"Did they acquire someone?" asks CampusCred's Sagar Shah.

"Yeah, it was on Hacker News. 'MongoHQ Acquires MongoMachine.'"

"Really? That's crazy. They're acquiring them while they're in YC?"

Someone mentions a rumor that a previous occupant of RocketSpace had rented a mansion and then rented out the rooms to other startups.

"Is this like a reality TV show?" asks Ballinger.

"That would be the most *boring* show ever," says Tamplin. He changes his voice, pretending to be the narrator. "And they're still typing!"

Attention is drawn to one of the friends of CampusCred, who is serving as a creative consultant for the T-shirts that they are planning to print up as a promotional item. The young man is introduced as "the sex writer for one of the Berkeley student papers, back in the day." Supposedly, this vocation led to his attracting all the willing sex partners he could possibly accommodate, though he himself remains silent as his alleged sexual escapades are narrated by a friend.

The guests listen to the tale with interest. Someone wonders aloud whether there was any way one could both do a startup and be a sex writer.

The conversation meanders on. Practical matters are taken up—founders exchange price information about large monitors that some are using—as well as financial ones. Even the founders themselves wonder aloud at how curious the system that has arisen is, in which investors entrust them with large sums of money in the form of loans that they will not have the means to repay if their startups fail.

"Do any of you guys know what happens to all these companies that die after the summer? What happens to the 150K that we receive?" Gutheim asks of no one in particular.

Narula doesn't understand his question. "'What happens?' What do you mean?"

Shah suggests, "No longer exists?"

"Yes."

"They probably spent it," says Pellow.

At the beginning of the batch, each startup used YC's forms to incorporate, creating a freestanding legal entity, of which they owned all but the 7 percent or so share that went to YC for its investment of $11,000 to $20,000. The Start Fund's and SV Angel's convertible notes were a $150,000 loan bestowed upon the corporation and not upon the founders personally. Because of the limited liability of the corporation, which shields the personal assets of its owners from the reach of the corporation's creditors, the loan entails no personal indebtedness on the part of the founders.

Gutheim, who is from Germany, is not clear about the legal obligations of the founders. "How does it work out? Do they get back to you and say, 'Hey, guys, we want our money back?'"

"No, no, no," says Shah. "If the company has no assets, there's nothing that they can do. If the company goes under, it goes under."

"Technically, it's still a debt—it's a loan, right?" asks Gutheim.

"It's the company's loan," says Narula.

"I got ya."

Narula expresses his own appreciation for the concept of limited liability: "God bless America!"

The conversation moves on to other topics: The 95 percent of YC startups in the previous batch rumored to have raised money from investors. The prevalence of startups in their own batch whose ideas pertain to advertising. Who present is using which database software or services: MongoDB, PostgreSQL, MySQL, Redis. The way that YC's winter 2011 startup LikeALittle, which calls itself a "flirting-facilitator platform," is apparently attracting users in abundant numbers. The way that same company has no business model in view.

Ballinger thinks LikeALittle is doing exactly what it should. "I think for consumer stuff, you get a bunch of users. Now that you have them, you figure out how to make money."

"The consumer Internet is a little different from what you and I do," says Narula.

"I think it's all about growth, right? It's like if you had a chance to invest in Facebook five years ago, before they had any clue about how to make money, it still would have been a good decision. Just based on growth."

Tamplin joins in. "Sift Science is growing really quickly!"

"But we're not consumer," says Ballinger.

"The next Facebook," Tamplin says, and Ballinger laughs.

"Do you have reven—?" asks Narula.

Before he can complete the word "revenue," Ballinger is answering, "No, no, not yet. We're working with our first set of customers. Get them happy with the results. Then—"

"That's what we think," says Narula. "We were thinking of charging people. Then thought it's better to at least prove that we work."

They all ask one another about future pricing plans and their products. CampusCred's Ben Pellow wants to know what kinds of work MobileWorks handles for its clients. Pulling information off Web pages is one category of task. An example: having mobile workers look at an online advertisement and assign a subject category.

"What is the cost breakdown?" Pellow asks.

"We charge per task, instead of charging hourly," says Narula.

"I kind of understand why you'd charge per task, but if you *were* to break it down, could you give the number?"

"Some tasks are more complicated so it's more time." He pauses. A typical price would be five cents per task.

"So that would be like per Facebook page that they visit and scrape something?"

Each piece of information would be treated as a separate task, Narula explains. Still, "it's cheaper than most of our competitors. The other aspect of it is it's high quality, socially aware."

Narula is asked when his last day off had been. "I took off Sunday three weekends ago." That had been his first day off in two months. He allows that he does not work every minute he is awake, every day. "Sometimes on Sunday I can't work for ten hours, twelve hours." Sleeping in on Sundays and working only six or seven hours is the closest he is getting to one day off a week. He sounds sheepish in confessing to the group his sloth.

Narula has been working seven days a week for a long time. Before YC began, he spent five days a week on school and two on the startup. "Two days—you couldn't get anything done in two days," he says. "And now I'm working seven days a week and even then I'm still running out of time." He must get a new visa, however, or he will have to return to school. "If I have to focus on school on top of that . . ."

It has been another long workday and work evening for everyone, but this one has ended amid the companionship of peers, a break from the routine. As fatigue sets in, founders pack up and leave in twos and threes, headed home.

13

NEW IDEAS

In the YC universe, business schools are deemed so useless that no one bothers to expend any energy in remarking upon their irrelevance to software startups. The only question relating to formal education that does draw attention is whether to finish or even go to college.

In 2010, Peter Thiel, the billionaire cofounder of PayPal, announced the establishment of a new fellowship that would award twenty $100,000 grants to entrepreneurs under the age of twenty if they left college to pursue their ideas.[1] As a debate spread across online forums about whether entrepreneurial young adults should bother with college, Justin Kan posted an essay on his personal blog in early 2011 about his own college experience at Yale.[2] He tries his best to come up with an itemized list of knowledge and skills acquired at college that he is now using as a technology entrepreneur, but the list is pretty meager.

Getting into Yale, he says, was the hardest part. He found he had to do less and less work for his courses. "Instead of studying, I spent my time: drinking and socializing; playing video games; working on a calendar website [Kiko]." He was appreciative of college as "a pretty decent club for me to hang out while I waited until I was ready to be an adult." He majored in physics and philosophy but was not taught the skills that he would need as a founder of a software startup. He itemizes the areas: programming, Web development, design, product management, project management, accounting, corporate strategy, business communication. College friends in other fields—science majors who went on to medical school or economics majors who landed at hedge funds—report that they too were poorly prepared.

Yale does not offer an undergraduate degree in business, but Kan is not critical of Yale's curriculum. What he most needed to learn—"to think outside the box, live the consequences of my own actions, or really exist on my own in the real world at all"—would not be found in any lecture hall.

Neither Emmett Shear, the computer science major, nor Kan, who picked up the rudiments of programming as they worked on Kiko their senior year, was interested in going to work for a tech company after graduation. They wanted to be startup founders, on their own. YC gave them the opportunity to be exactly that. Then, at the end of the inaugural YC session that summer, they raised $50,000 from investors to push on with their calendar Web site.

This first startup ran aground, however, when Google released a Web calendar similar to Kiko. They decided in 2006 to get what they could for their Kiko source code on eBay, hoping to repay their investors, and ended up with far more, netting $258,100.[3] Despite the end of Kiko, the experience left them as committed as ever to startup life. They now needed to come up with a new idea for the next startup.

Kan and Shear were spending time with the founders of YC's summer 2006 batch and talking about idea possibilities. Returning after a YC dinner, Kan came up with a new idea: how about providing on the Web a live audio feed of his and Shear's discussions about strategy? Might that not be of interest to others who were entrepreneurial minded? Wait—even better: a live *video* feed. And why restrict it to certain conversations? What about a continuous live video feed showing one person's life around the clock? And add chat, so viewers could talk with the subject and among themselves as they watched. Kan was willing to be that person and came up with a name: Justin.tv. Paul Graham had YC invest and though Kan and Shear did not formally belong to a YC batch, they became de facto second-timers.

When they moved to San Francisco in October 2006 to begin work on Justin.tv, they had no idea what the cost of streaming video would be or how they would acquire customers or secure sponsors or sell advertising. They were oblivious to the challenges they would face making their own hardware that could send video over the limited data connection of cell phones back then (pre-iPhone). They did not have a scintilla of knowledge about what kinds of live video would appeal to viewers. They were going to enter a hits business without any experience making a hit. Looking back years later on all that they did not know, Kan would title a blog post "Why Starting Justin.tv Was a Really Bad Idea but I'm Glad We Did It Anyway."[4]

Not everyone would be willing to attach a video camera to his head and broadcast everything that happens. When Emmett Shear is asked, years later, if there could have been an Emmett.tv, he laughs at the absurdity of the question: there was no way he would have done what Kan did. It was precisely because it was so unappealing to most people that the startup attracted lots of press attention when Justin.tv began airing in March 2007. In the meantime, Kan and Shear had added two other co-founders: Michael Seibel, a fellow Yalie, and Kyle Vogt, who left MIT and moved to San Francisco to join them.

The founders did not have anything resembling a plan, other than to start and see how much interest Justin.tv would attract. About two thousand people became regular viewers, making Justin Kan a minor celebrity who was recognized on the streets of San Francisco. But that was not a base sufficient to support even a tiny four-person business—and Justin.tv had neither sponsors nor advertising. The founders sent out video hardware to individuals whose lives seemed interesting, hoping to get them to broadcast continuously, too. But the results were disappointing.

"How much goes into producing a cable news channel? Way more than one guy walking around with a camera," Kan says, laughing as he looks back.

Shear adds: "To produce one hour a week of good TV—even mildly terrible TV—takes a significant number of people and a lot of effort. One person can't produce actually interesting content all the time."

So they decided to open up the site to anyone and everyone, making it a general-purpose platform for live video. They had launched the site without realizing how costly it would be to send out video. In their first month, they received a bill for $38,000 from their content delivery service—when they had only $15,000 in the bank. But they were adept at raising money: a few hundred thousand dollars from angels immediately; then, in 2008 and again in 2009, they raised a total of $7.2 million, as well as securing a $2 million bank loan.

Many prospective investors remained wary of venturing anywhere near what appeared to be an entertainment business that would be based on hits. But Kan and Shear were passionate—as Kan put it in his blog reflections, "We sold the shit out of it"—and their enthusiasm proved infectious. Raising so much capital gave Justin.tv's founders a lot of time to learn the things that they would need to know. And they express no regret that they had not taken business courses in college. Learning what you need to run a business, when you need to, is part of the appeal of startup life.

When Justin.tv opened to everyone, traffic grew and grew—sometimes by 50 percent in a single month. The founders had to figure out how to build their own network infrastructure to reduce their costs by two-thirds, which they did. They began running advertising in 2008 and Justin.tv did well enough to survive. Some thirty million people—"uniques," as they're called in the Web business—spent time at their site (the person who comes to a Web site on two different days in the month still counts as only one—i.e., unique—person). Considering the bandwidth-intensive nature of video, by 2010 the company may have been moving more bits than any other Web site in the world that is not a household name.

Growth slowed, however, and when it did, it became difficult to retain employees. "When you're winning a lot, when you're tripling every year, or you're 10x-ing every year and you're already big, nobody's going to leave that because it's obviously a winner," says Shear. "But when you stop growing a lot, retention and motivation for people becomes a much bigger problem. You can't motivate people just by, 'Oh, we're on this rocket ship of awesomeness.' "

In fall 2010 the four Justin.tv founders decided the company should consider moving into a more promising business. Looking back on that moment, Kan says, "It was pretty clear that the business was not going up. We wanted to be at the next level or just go broke. None of us wanted to work at a company where we pay ourselves a very nice salary for fifteen years but the company wouldn't be worth a lot."

Shear adds, "In the Internet world, you're never even guaranteed that. The sand shifts beneath you. Unless you get really, really big, medium-sized Internet companies have a tendency to have their market eaten by someone else who is much bigger."

The startups who go through YC typically start with a modest idea and then Paul Graham provides a way to think about expanding it into something bigger. What Justin.tv's founders were considering was the opposite course: abandoning the general-purpose platform and growing by focusing on only one small piece of what they already had.

Shear wanted to focus on gaming. He had looked into the business and had seen that it was enormous—it included not just companies publishing games but also media businesses that catered to gamers. Online videos for gamers produced by a single company, Machinima, were attracting billions of views annually. Gaming seemed to Shear a natural choice: Justin.tv already had a section for live streaming of gaming and it was the one category of content on the site that he himself watched.

Michael Seibel, however, wanted to work with videos recorded with smartphones. Justin.tv had a broadcasting app, but it was not used much. Perhaps a better approach would be to replace mobile live broadcasting with a smartphone app that would make it easy to share videos once they had been shot.

Which was the better idea? It was not clear. The cofounders decided to create teams to work on each one, set a month-over-month sales goal to gauge user interest, and see which one hit its milestone. If one succeeded, that would be the answer. If neither did, they would discard both and start over. And if both hit their milestones, that would be a nice problem to have.

Shear led a four-person team to work on gaming. Kan led another four-person team to develop a new video-sharing app, which would be named Socialcam. The teams gave themselves six months to see what could be learned about the potential of both ideas.

The place dedicated to live gaming that Shear set about creating was initially hosted at Justin.tv. Gamers could broadcast the games they played on their PCs; spectators could watch and chat online. Shear also prepared to launch a separate site that would have an identity that was unmistakably associated with gaming: Twitch.tv.

Gaming was one of the largest categories of video at YouTube. Some of the YouTube videos were recordings of gamers playing. Twitch.tv would offer the immediacy of live competition. Other startups had tried to do something like this, but technical problems had afflicted the earlier efforts. The back-end infrastructure had to be able to support the load and Internet connections at home had to be fast enough to reliably support high-resolution video. If Justin.tv had chosen gaming as its core idea four years earlier, it could not have succeeded: its infrastructure would not have been able to meet the demands of gaming, nor would users have had the fast Internet connections needed for a pleasing experience. But in early 2011, the technical requirements could be met.

The timing was propitious for another reason: the business model that game companies had relied upon in the past was crumbling, making them eager to work with a Web site that attracted their customers. "Gaming companies are very aware that their current model of selling fifty-dollar games in boxes in stores is not long for this world," says Shear. "People are still willing to pay money for games, but they're all looking at games like World of Warcraft, which you get for free or for very cheap, and then you pay incremental amounts of money to unlock additional functionality." The companies are asking, "How do we continue to engage someone in the game as long as possible?" Zynga, the publisher of FarmVille, Words with Friends, Mafia Wars, and other free or nominally priced Facebook and iPhone games, is an example of the new form of game company. It depends upon its users remaining engaged with the game in order to generate revenue.

Publishers of older games that originated on game consoles managed the games like a series franchise. "We're effectively on Call of Duty 10 now," says Shear. "The publisher's worst fear is you'll start playing a different shooter before the next one comes out. So anything they can do to keep you watching and engaging with Call of Duty until Call of Duty 11 comes out is huge. It keeps mindshare." Twitch.tv would try to attract the best Call of Duty players in the world, as well as the best players for World of Warcraft, League of Legends, and all the others that had made online gaming tournaments a spectator sport with enormous followings invisible to those who did not share the gamers' interests.

Few among the most highly skilled players could support themselves by playing in e-sports tournaments, where a single winner typically received all of the prize money. Shear, however, was offering skilled players the opportunity to become professional players by coming to Justin.tv and streaming their play. It was easy to do: the players would boot up the game on their PC, turn on video capture software, and start streaming to the site. Players had been invited to stream before this, but they had to serve as a commentator, too, narrating their own play. This entailed hooking up a microphone and diverting a portion of their attention to speaking instead of focusing fully on the play. Now Shear was telling these players, "Let's just try streaming alone." For players who drew a significant number of fans, Shear offered a fifty-fifty split of the advertising revenue their broadcasts brought in.

The experimental expansion of Justin.tv's gaming section showed encouraging results immediately. Shear and his team prepared to open the separate site for gamers, and in May Twitch.tv was opened on a limited basis to invitees only. In June, the site was opened to the general public.

Tyler "Ninja" Blevins, one of the star Halo players that Twitch.tv brought on, found he could make about a hundred dollars a day from his share of advertising revenue. His "workday"—gaming—was ten hours but he said he did not mind at all: "I love playing. I've never, never not enjoyed playing Halo."[5]

Bringing more live feeds of the best gamers showing their skills brought yet more game enthusiasts. And once at the site, they watched for extended periods of time. Live streams of gamers would keep fans riveted in a way that grazing on short YouTube clips did not. Over a month, Twitch.tv drew more than three million unique visitors; each one stuck around to watch an average of four and a half hours of live gaming competition. These numbers immediately caught the attention of game publishers.

While the gaming initiative progressed, so too did the work on mobile video sharing. In March, at the South by Southwest conference in Austin, Justin Kan and Michael Seibel introduced the Socialcam app for both iPhone and Android. After an updated version was released the next month, the app facilitated video sharing with Facebook, Twitter, Posterous, Tumblr, and Dropbox, or via e-mail or text messaging. The company said the app was downloaded 250,000 times in its first month, though it did not say how many users actually used it regularly. When a TechCrunch reporter asked what subjects Socialcam users were recording and sharing, the reply was, "Lots of pets and babies."[6]

It's June, six months after the dual experiments began, and Justin.tv faces the problem the founders were hoping to have: choosing between two initiatives, Twitch.tv or Socialcam, that have met their assigned milestones. For the moment, the founders decide to invest in both and let the experiments continue to run.

First Kiko. Then Justin.tv. Now the two new ideas. Kan and Shear have had perhaps more experience taking new ideas to Paul Graham than any other pair of founders in YC's history. They credit him for supplying periodic boosts to their spirits when their own motivation has flagged. Kan says:

> Very few people, I think, have an energy like that. You can read about customer development online or watch Eric Ries talk or look at Dave McClure's AARRR framework for startups.[7] That's all great. But when you have an in-person conversation with Paul, he might say some of your ideas are shitty—he might say *all* of your ideas are shit—but when you walk away, you want to go build something.

✦

Zach Sims and Ryan Bubinski, the duo trying to get BizPress aloft, have found the going difficult. Bubinski has written code that, theoretically, makes it possible for a small-business owner to quickly design a customized Web site by dragging and dropping some widgets on the computer screen. But they've discovered that the typical business owner is not sufficiently computer-savvy to feel comfortable even with dragging and dropping. For those businesses that do have a Web site of their own, the site may not have been updated in the years since a high school kid was paid a few hundred bucks to create it, and the owners have little interest in giving the matter their attention now. The two wonder: "Is this how we want to spend the next ten years?"

In late July, they decided to take a break from BizPress and do a weekend project, creating a smartphone app, Thing Marks, that Sims thought might be a good alternative idea for their startup. The app would present a map upon which the user would mark places, like a restaurant or bar, to be remembered for future occasions. They built it over the weekend and showed it to Paul Graham. It's a terrible idea for a company, he said, and they lost their enthusiasm for pursuing it. But they discovered when they returned to working on BizPress that they could not muster enthusiasm for it, either. With only one-third of the summer remaining, they did not have much time to come up with something else.

In these unlikely circumstances, the idea for Codecademy appeared. During the summer, in the evenings, Sims had been trying to learn more Ruby so he could help code alongside Bubinski. But reading books and watching online instructional videos was painful and had left him frustrated. It was like reading how to play basketball for two hours and then playing for five minutes. Perhaps he and Bubinski could create an interactive site for teaching nonprogrammers like Sims how to code, giving students the chance to learn bite-sized concepts and then immediately try them. They also thought of the nontechnical people they knew in New York, like investment bankers and consultants, who were constantly talking about their interest in doing a startup. These nonhackers said they had ideas for startups and were searching for technical cofounders, convinced that learning how to code was too difficult to undertake themselves.

They took this idea to Graham, who liked it. He thought they should make the programming lessons into a game. Even though the two had started the summer originally thinking they would use games for programming challenges, they were now cool to this suggestion. They wanted the courses to appeal to everyone—so no games.

They began working on the first course, which would teach JavaScript. Bubinski had long lost the ability to view a programming language through the eyes of a novice programmer, so it fell to Sims to read a pile of books on the subject, consult with Bubinski, and then write the lessons. Starting when they did, with Demo Day only four weeks away, they could see that it was not likely the site would be ready in time. They asked Graham if they could skip Demo Day. They knew it would be embarrassing not to present, but less embarrassing than presenting a Web site that was egregiously incomplete. Graham suggested they wait and see whether embarrassment was certain before deciding.

Codecademy is the third idea that the pair is trying out this summer. It is a shame they did not start with it and have three months to work on it instead of one. The additional time would have been put to good use. They have spent the summer working in the Spartan apartment in Sunnyvale, seven days a week. But not every startup gets to follow a straight flight path upward. The founders of Codecademy will not, in the end, have any reason to be regretful about how things turned out.

14

RISK

YC partners are interchangeable, Paul Graham told the founders at the beginning of the summer. You can come to any of us for office hours and receive the same advice. In practice, this applied only to the basic YC tenets: work on code and talk with customers; launch fast and iterate; focus on one measurable weekly goal. The founders quickly saw that, beyond the basics, there was plenty of variation. Like an academic department, YC partners were distributed across generations. Graham came of age before Paul Buchheit, who came of age before Harj Taggar, who came of age before Sam Altman. Additional coverage was provided by the other partners: Justin Kan, Emmett Shear, Garry Tan, and, near the end of the summer, the newest partner, Aaron Iba. Founders sense that some partners like their startup's idea more than do others, and naturally gravitate to them.

Graham is viewed as the best person to go to when thinking about how to expand the idea, and he is also the inspirational figure who can leave founders feeling a surge of energy after office hours with him, eager to get back to their apartment and resume coding. But for help on product design issues, many founders view the advice of the other partners as more reliable. They also know that Graham's blunt criticism can feel crushing. His colleagues are never immoderate in what they say. When they are critical of the founders' ideas, they soften the edge of their criticism, claiming that they, the advisers, may well be wrong.

In YC's first six years, the format for office hours did not change.

Founders met with one partner at a time. If they wanted multiple perspectives, they set up office hours with each partner separately. Depending on the partners' schedules, founders might have to come in on different days.

This summer, however, Graham is trying out two new formats. One is group office hours—two YC partners meet with the founders of six startups, giving the founders the opportunity to see how the partners respond to the questions posed by other founders as well as their own. The other experiment is called, inelegantly, "inflection point meetings," in which the founders of one startup come in and get the undivided attention of a full panel of YC partners. This cannot be offered to every one of the sixty-three startups; it is reserved for those deemed to be at a critical juncture and in need of the advice of more than one partner. The format suggests that the partners are not fully interchangeable after all and that startups benefit by exposure to many points of view.

It's a Monday afternoon in early August, and the three founders of Tagstand—Kulveer Taggar, Srini Panguluri, and Omar Seyal—have been invited in for an inflection point meeting. Though Taggar's surname coincidentally bears a passing resemblance to Tagstand's name, the startup's name derives from the tags it sells, each containing an NFC chip. Eventually, NFC technology is expected to enable paying for purchases with a wave of the phone, replacing credit cards. Tagstand is trying to make a business selling tags to businesses that are trying out NFC for purposes other than payments, such as in advertising. For example, a "smart poster" with an NFC tag affixed to its corner would allow a curious passerby to bring up on her smartphone a Web page related to the advertisement when she places her phone by the tag.

The Tagstand founders come into the small meeting room and take their seats. Gathered around the table are Graham, Buchheit, Tan, and Iba. Kulveer's cousin Harj is not present, but an unofficial, honorary partner, Geoff Ralston, is sitting in. Ralston is an angel investor and friend and contemporary of Graham's. In the 1990s, he cofounded Four11, a startup that was acquired by Yahoo in 1997; Four11's product, RocketMail, was rebranded as Yahoo Mail.[1] Ralston worked at Yahoo as a senior executive and later went on to found another startup, Lala Media, a streaming music service, which was sold to Apple in 2009.[2] He has invested in a number of YC startups and has just cofounded Imagine K12, a YC-like program for startups that develop software for schools.[3] In recent weeks, he has been spending time at YC and offering office hours to founders.[4] Like Graham, he is not shy about expressing his opinions.

Graham begins: "So, you guys. I feel like I haven't talked to you guys for a while. I don't understand what's going on. So what *is* going on? You changed the plan, right? In the course of YC? From being a supplier?"

"To being a general platform, a Heroku type of thing," says Taggar. "But we realized it's too early for that. So now we're going to focus on the vertical of outdoor advertising."

"So that's *another* change. What was the first thing?"

"The first change was we were just a store," selling NFC tags to publishers and merchants who use them in their marketing. Taggar continues, "That was just a way to get in touch with the early adopters. And it actually worked. Our revenues have doubled every month. We've got a whole lot of leads that way, companies like ConAgra and Topps trading cards and Condé Nast. So the whole framing of us as NFC experts is kind of working. And we're just getting loads of inbound sorts of inquiries."

Using a Web address that Taggar has supplied him, Graham peers at his laptop, looking at data that Tagstand maintains about the use of its tags. "So the most taps, ever, on any one of your customers is only eighty-two?"

"Yeah."

"That's bad."

It's not surprising, though. NFC-enabled smartphones are scarce in the United States. Currently, only one smartphone model at one wireless carrier has NFC.

"Are you sure it's going to end up in all of them?" Graham asks.

"Yeah. That we're sure of."

"So what you have to do"—Graham chuckles—"is survive until anyone cares about NFC."

"So this is what we were thinking about," Taggar says. "This whole outdoor advertising angle? We wouldn't just be NFC. It would be QR codes and SMS and near-field. This came up in our conversation with Sequoia." A partner with the venture capital firm had told them, "There's no value in NFC yet because no killer application has been done. So why don't you guys get a killer application?"

They had drawn up a list of the most promising possibilities. Leading the list was payments, but it was already crowded with the likes of Visa, Google, and other heavyweights. Using NFC for ticketing was another possibility—with a wave, an e-ticket could be purchased and downloaded to a smartphone—but only in the future, when more users had NFC phones. This had left the residual choice of outdoor advertising.

Taggar describes their intention to buy advertising space at bus stops in San Francisco and then offer it to YC companies like Dropbox and Airbnb.

Graham has not heard of this plan. "No shit? Bus stop advertising!"

"On the guerrilla side of that, we've actually printed up these labels, which are like, 'Tap here' or 'Scan a QR code.' We're going to put them up on MUNI stops around San Francisco. Just to see what people do."

Geoff Ralston speaks up. "Aren't a lot of people doing that kind of thing?" He knows that there is a startup in this batch that is doing QR codes—Paperlinks. "Isn't that really a crowded space? Or becoming one?"

Graham tries to turn the conversation back to the question: how is the company going to generate revenue? "You've got to figure out some plan for making money so you can survive."

Ralston speaks for Tagstand: "Selling tags."

That may work, Graham says. "You're selling shovels to the miners, and the miners aren't making any money yet. That's OK—you are," says Graham. "What *is* your revenue from selling tags now?"

It was $1,200 the month before, but it looks like this month's revenue is going to triple. Their costs are low: they purchase the tag for thirty cents, pay piece-rate workers found on Craigslist ten to twenty cents to program it, and then sell it to companies doing experimental marketing with NFC for one or two dollars.

Ralston does not think much of their plans to focus on outdoor advertising. He suggests that they instead think about how to go about preparing to be the one company that others think of as "the NFC guys." He says, "If you can build that brand, there's probably some pretty significant motivation for some folks to work this stuff out. Just a couple million dollars from people that will let you survive for four years. In the meantime you're building your platform on it, so by the time NFC starts to happen, there's no other choice. You guys have figured *everything* out. You know exactly what to build. When the first real deployment comes, no one can really compete with you."

This is what Ralston calls the "optimistic" scenario. "The pessimistic scenario is, because you're so early, you get all this shit wrong. And it turns out in a year you can build everything you need to build. And someone comes in three years from now, figures that out, and actually builds it better because you guys have all sorts of legacy shit. Then you get beat. So that's the risk."

Taggar says Tagstand has raised $200,000 from the Start Fund and SV Angel in addition to the $150,000 that all the startups received at the beginning of the summer.

Graham was unaware of this. "You got more money from both?"

"Yeah," says Taggar. "Probably that's partly the team, because we're second-timers"—a reference to his first YC batch, winter 2007, and Panguluri's first, in winter 2006. "But we're finding there's basically investors that are really gung ho that NFC is going to change the world. And then there are some that don't buy it." He feels confident that there are enough investors who are enthusiastic that Tagstand will be able to raise a substantial amount of money.

"Maybe there's some revenues in two years, right?" suggests Ralston. "There's like limited deployments—for specific reasons, like the U.S. Army has NFC for something and they need you and will pay you $10 million and you can get real revenues."

"The other thing that we could do—we've talked about before—is the whole international angle," says Taggar. "I just found out too late. In Singapore, all taxis, already—you pay with NFCs in them. And in Japan, there's like seventeen million NFC phones," though these were the "dumb" kind that came before smartphones. Europe is also far ahead of the United

States in deploying NFC-equipped phones. He adds, "The idea of basically trying to survive for four years does not appeal to me."

"You mean you want to be successful, not just survive?" asks Graham.

"Yeah!"

"Then why did you choose NFC?" asks Ralston.

"We wanted to learn about it."

"I guess we've done that," Panguluri says dryly.

Taggar points to major deployments for Tagstand that may be coming up quickly. "If we get this Google thing, and Condé Nast wanted three million, there is a scenario whereby—"

"Condé Nast wanted three million?" asks Garry Tan.

"Yeah."

"A dollar each?"

"We're not sure they really understand the costs."

Several partners express doubts that Condé Nast is about to spend $3 million to insert tags into their magazines that their readers' phones cannot use.

Ralston says, "It sounds like in certain markets—seventeen million in Japan—is there already a platform business there? What are they doing? Maybe you guys just need to be an overseas company for four years. Instead of surviving, you become a $150 million company in four years there and become a $500 million company once it actually rolls out in the United States when we get our act together. If there's millions of these things being manufactured, someone's using them somewhere—someone's buying them. For expensive prices, too."

Graham speaks up. "Fundamentally, the reason this startup is so confusing is because the situation is confused. That's what's really going on here. The danger when you're talking to investors—I find Tagstand confusing. What's 'Tagstand' mean? Something to do with NFC. That *is* in fact what you're doing: 'something to do with NFC.' You're not sure either. You've got to have a clearer story for investors. The story's got to be: 'NFC's a big deal. We're not sure how yet. But we want to make sure we're the people who figure it out. And the way we're making sure—and also not running out of money—is we're selling the tags to people so that way we know everybody who's using these things. And we have the leads when we want to sell them something. And we're also going to do X, Y, and Z.' The question is: what *is* X, Y, and Z?"

On his laptop, Graham is looking at Tagstand's Web site. Tagstand's "NFC hobbyist starter kit" has caught his eye. "My twelve-year-old self goes, 'Wooo! What can I do with this?'" His voice is excited. "It reminds me of when my twelve-year-old self was looking at the Commodore PET. Microcomputers started out as this thing for hobbyists. You guys should be the suppliers to all the hobbyists. Maybe if you get more and more into the business of selling people stuff, that could actually be the answer."

"Like Tamagotchi," suggests Buchheit. "Where you have to touch the Tamagotchi to your three friends' phones or something—"

Graham asks the founders, "How close do you have to get to an NFC tag to set it off?"

"Ten centimeters," says Taggar. "It depends on the size of the thing."

Ralston reviews Tagstand's strategic options. Move to outdoor advertising and QR codes. Go international. Or simply hunker down and try to survive doing what they are currently doing, selling small lots of tags used as experiments by marketers. He has a cautionary tale to tell about the risks of waiting years for NFC's moment to arrive. He mentions a company that had believed in the promise of OpenID, a technology that had the potential, if widely adopted, to allow a user to supply the same user name and password to log onto all Web sites. The company spent five years building infrastructure to handle OpenID. The story does not have a happy ending. "The problem is OpenID didn't become anything—Facebook killed it. When you waste four years for a technology—technology changes pretty fast—and NFC might be irrelevant."

"NFC seems like it's going to be something real, more so than maybe QR codes," says Buchheit. "Because it's useful for payments or whatever."

"I guess so," says Ralston. "The thing about QR codes is I'm starting to see them in places that are really surprising to me."

"They're all over the place," Buchheit agrees. "But I've never seen someone use them. I think they still might be a fad. Because it takes effort." NFC seems to him more practical, as he imagines using it to pay for things or unlock the door of his house with a wave of his cell phone.

"QR codes are *so* much cheaper than these things. So much cheaper," says Graham.

Ralston is unconvinced that tapping an NFC tag is considerably easier than scanning a QR code.

Not so, says Buchheit. NFC is nothing but a tap. But a QR code requires several steps: "Let me unlock my phone, launch the app that does that, line it up—it's a completely different product in my mind. Because that speed bump is so huge," he says.

Taggar directly challenges the notion that Tagstand will have to bide its time for four years: "The premise that there can only be an NFC business when 50 percent of smartphones have it is kind of wrong." The Sequoia partner had suggested to him that once 10 to 20 percent of phones are NFC equipped, NFC will have passed the critical threshold. "It's not going to go away. There is a scenario where, by this time next year, basically all the new Android phones have this technology, all the new BlackBerrys have it. HTC told us that all of their phones next year are going to have NFC. So it's not necessarily now or four years. It might be a year."

This does sound promising. The partners return to QR codes and outdoor advertising, the direction in which Tagstand had said a few minutes earlier that it wants to head.

"Remember the CueCat?" Buchheit asks the group. He laughs. "This was like a nineties thing where they mailed you a bar code scanner. And you were supposed to scan the bar codes in all of the advertisements in *Wired* magazine. You would plug a scanner into the serial port of your computer, install software, and then use it to scan the bar codes on the advertisements." The scan would bring up a Web page on the PC with additional product information—essentially, a user was asked to go to considerable trouble just to get more advertising.[5] Thinking about it, he laughs again. "There were like three steps of implausibility. And it raised a ton of money because they had to mail everyone a bar code scanner. I got one. They just sent them to every subscriber of *Wired* magazine."

Ralston remembers it. "You got this and you said, 'This is extraordinarily stupid.' You knew it immediately." He does not feel the same about scanning QR codes, though.

"Oh, no, no," says Buchheit, making clear he did not mean to imply that QR codes are the same as the CueCat. "But it's that same dream, 'Oh, this is great, 'cause our customers want to interact with advertisements.' I think the answer is, 'Not really.' I think marketers are deluding themselves about the extent to which their customers want to interact with the advertisements."

Ralston gets his phone out to see how long it takes him to scan a QR code. He succeeds quickly.

"That was surprising," Graham says, impressed.

Buchheit tries it himself but does not succeed as quickly.

Whether the QR code is easy to scan or not, Ralston does not want to see Tagstand pursue QR codes and outdoor advertising simply because the founders regard it as a decent business opportunity for the moment. They should be "passionate" about the business, he says. "The best people really love the business they're in. They do. They love it! Right? And I think if you love *this*"—indicating NFC—"and you really believe it's a big business, I don't want to dissuade you from going after your true love."

"Ehhh." Graham does not agree with Ralston. "You can't love online store software enough, you know?" says Graham, a reference to how he and the Viaweb team did quite nicely when Yahoo had acquired their company, which was devoted to online store software, hardly the stuff about which he or his colleagues had ever felt passionate.

"Right," Ralston says, realizing that Viaweb does seem like the perfect counterexample to the point he has just been making. He turns to Graham. "You loved it enough. I mean, you sold it to a schmuck company, got lucky, right?" Ralston can tease Graham about selling to the schmuck company and getting lucky because Ralston had sold his startup to Yahoo, too, and stayed, working as an executive for the schmuck company for far longer than Graham had. If Yahoo had not bought Viaweb, Ralston tells Graham, "you probably would have fallen out of love with it and done something else." He laughs.

"I don't think any of us have this innate love of NFC technology," says

Taggar. "But I think what *is* exciting is the feeling that we *are* on the cutting edge and pretty much anybody that wants to do stuff with this, they do come to us. And, yes, we know we're early. But it does feel like, yeah, this could actually be quite revolutionary."

"Keep growing your revenues and do things that make you important," says Graham.

"It may take a long time, though," says Buchheit. "Microsoft was founded in 1975 and probably no one ever heard about it until the eighties."

"Things go faster now, though," says Graham.

Buchheit says that with NFC designed into all Android phones, its diffusion seems inevitable.

Addressing the founders, Graham says, "You can be candid with investors and say, 'We don't know what we're doing. Something is going to happen with NFC. We're going to make sure that we're in a position to be the ones who find it.' "

✦

The Tagstand founders leave with the tacit encouragement of the YC partners to wait and see what opportunities NFC opens up. This is unusual. The bedrock advice in the YC User's Manual is to make something people want. These four words—"Make Something People Want"—were emblazoned on the T-shirts printed up for the first YC batch in summer 2005, and this has remained the YC mantra.[6] Kulveer Taggar made something that people wanted in his first YC startup, Auctomatic—not something that he himself wanted, but what eBay's largest sellers wanted, software for uploading and managing eBay listings in bulk. This time, however, he shows no interest in repeating what he had done before, identifying a visible need and then addressing it. He and his Tagstand cofounders are deliberately moving ahead of what can be known about what users will or will not want.

Graham approves of the Tagstand founders' staking a claim where they have because he believes that rewards are closely linked to risk. "The way to get really big returns is to do things that seem crazy," he wrote in 2007, "like starting a new search engine in 1998, or turning down a billion-dollar acquisition offer"—a reference to Facebook.[7]

It is investors, however, not founders, who are best positioned to seek outsized rewards by taking on outsized risks—investors have the protection of a portfolio that spreads the risk. Graham understands that investors like YC will naturally have a greater appetite for risk than individual founders, and he has tried to come up with a way to close the gap by encouraging founders to serially start lots of startups. A succession of startups would spread risk across many bets. The problem with this argument, however, is that startup founders cannot start all that many. Even those in the summer batch who had been in YC earlier—Taggar and Panguluri in Tagstand, plus some other founders in Parse and Can't Wait—were only on their second ones. Two, or even three, startups does not generate much of a portfolio effect for founders. The fact that YC's summer batch consisted of sixty-three startups suggests that a portfolio needs dozens of seed investments, not singletons, to work as it should.

No statistical table exists showing that rewards for startup founders are neatly, predictably matched to risk. Founders who have chosen an idea that is highly risky, in the hopes of reaping commensurate rewards, are arguably guided more by emotion than anything else. This does not make it the wrong decision for the founders. But they should keep in mind that Graham, the investor, has a bigger appetite for risk than did Graham, the startup founder. Before he started Y Combinator, he wrote:

> Viaweb's hackers were all extremely risk-averse. If there had been some way just to work super hard and get paid for it, without having a lottery mixed in, we would have been delighted. We would have much preferred a 100% chance of $1 million to a 20% chance of $10 million, even though theoretically the second is worth twice as much. Unfortunately, there is not currently any space in the business world where you can get the first deal.[8]

15

MARRIED

When Y Combinator began, Paul Graham set down a formal rule: no funding of startups with only one founder. He later relented a bit, allowing occasional exceptions in cases where the founder seemed extraordinary. It was a good thing he did relent—one of those exceptions was Drew Houston, the sole founder of YC's largest success: Dropbox. But Houston did not remain the sole founder—Justin.tv's Kyle Vogt introduced him to Arash Ferdowsi, who became a cofounder. If anyone doubted the need for at least two founders, Graham would list the technology giants whose public identity is bound to a single founder—Microsoft (Gates), Apple (Jobs), Oracle (Ellison)—and then point out that each one initially had two founders, not one. All startups are the same, he maintains. You need two people to spread the load.[1]

Graham could enumerate other reasons for not wanting to fund single founders. The very fact that a founder did not have cofounders was evidence, in his view, that the founder's friends had withheld placing their confidence in the founder. "That's pretty alarming," Graham said in 2006, "because his friends are the ones that know him best." Even if friends had underestimated the founder's ability to make a success out of a startup, and even if the founder could handle all of the startup's tasks alone, the founder still needed others, Graham said, "colleagues to brainstorm with, to talk you out of stupid decisions, and to cheer you up when things go wrong."[2]

More than one founder was essential, but not many more. Harj Taggar, addressing aspiring founders in London in 2011, said that YC's experiences showed that four founders was too many—decision making was too cumbersome. "You don't want your startup to be like the UN, some sort of democratic process to get the smallest thing done."[3]

A startup could change its idea easily, but not its cofounders.[4] It was critically important that the two or three founders knew and liked each other well before starting out. Graham said, "Startups do to the relationship between the founders what a dog does to a sock; if it can be pulled apart, it will be."[5]

In 2009, Graham conducted a survey of the founders in YC's previous batches and was struck by the way that those in the successful startups talked about the intensity of the founders' relationship. He quoted one respondent:

> One thing that surprised me is how the relationship of startup founders goes from a friendship to a marriage. My relationship with my cofounder went from just being friends to seeing each other all the time, fretting over the finances and cleaning up shit. And the startup was our baby. I summed it up once like this: "It's like we're married, but we're not fucking."[6]

The question that Graham received most often from aspiring entrepreneurs was, "Where can I find a cofounder?" If the person asking it was still in college, Graham had a ready answer: look among your close college friends. He pointed out that school was where the cofounders of Google, Yahoo, and Microsoft had met. "Most students don't realize how rich they are in the scarcest ingredient in startups—cofounders," he told MIT students in 2006. "If you wait too long, you may find that your friends are now involved in some project they don't want to abandon. The better they are, the more likely this is to happen."[7]

Justin Kan and Emmett Shear, the part-time partners who had been founders in YC's first batch, were exemplars of close friends who worked well together. "They can practically read one another's minds," Graham wrote in 2010. "I'm sure they argue, like all founders, but I have never once sensed any unresolved tension between them." Of course, Kan and Shear did get an earlier start on building trust than most founders—they had been friends since second grade.[8]

✦

Among the founders in the summer 2011 batch, there are many pairs of cofounders or trios who seem to work well together as if they too have been friends since elementary school. In one case, the two founders are close indeed: Elizabeth Iorns and Dan Knox, of Science Exchange, are married.

Most YC founders are living together in the same space in which they work. It is mid-June and the Ridejoys—Kalvin Wang, Randy Pang, and Jason Shen—are living in a three-bedroom apartment in San Francisco. The common area serves as Ridejoy's world headquarters.

The threesome originated in a friendship that Wang and Shen struck up when they were students at Stanford. Neither one of them knew Pang then. Yet by summer 2011, who knew whom first is undetectable, and Shen would write about the three of them: "We're smarter/more cohesive than any two of us alone." How Pang got to know the other two is a tale of how founders can get to know another person well even if they haven't gone to college together.

In the fall of 2009, Shen and Wang got an apartment in San Francisco but didn't then work together. Shen, who had studied biology and philosophy, worked after graduation as the business manager of the *Stanford Daily* and commuted down to Stanford. Wang, a computer science major, worked at a number of startups and landed at Virgance, an incubator located in the South of Market area of San Francisco, not far from their apartment. The two needed to find someone for the third bedroom in their apartment.

They ran an ad on Craigslist, brought in a roommate whom they had to ask to leave after only three months, and found themselves having to start over. This time, however, they determined to go about their search differently. Shen convinced Wang that they should provide prospective renters with an abundance of information about themselves. In a blog post, Shen later wrote, "It turns out, being very explicit about who you are, what you're like and who you're looking for is a great thing."[9] The two set up their own Web site, JasonAndKalvin.com, and described in great detail their work, their interests, and the characteristics of ideal roommates.

Among the respondents was Randy Pang, who had graduated from Berkeley the previous spring with a major in electrical engineering and computer science, and who publicly shared *his* personal interests at his own Web site, RandyPang.com. After Pang was invited to move in, he became close to both Shen and Wang. (Two years later, when Wang teased Shen about his self-promotional blog persona and use of social media, he said he had to admit that providing what he personally regarded to be "ridiculously detailed" information certainly worked well when searching for a roommate. The title of his blog post: "How Borderline Douchebaggery Helps You Land a Great Roommate.")[10]

✦

Graffiti Labs is another startup in the summer batch that originated with a friendship in college, but it departs from the Viaweb model of two-hackers-who-meet-at-school. Tim Suzman and Mark Kantor, who are twenty-seven years old, became best friends as freshmen at Brandeis. Like Kalvin Wang and Jason Shen, this is a pairing of a hacker, Suzman, with a nonhacker, Kantor.

They have a third cofounder, whom Tim has known a very long time: his brother Ted, an ace hacker who is three years younger. The three worked on businesses of one kind or another while Ted Suzman was still in high school. A few years after his older brother and Kantor graduated from college, Ted left Washington University just short of graduation to work with them full-time.

In their YC application, when asked to provide an example of their most successful hack of a noncomputer system, Kantor said he had met Puff Daddy and got a job with the concert company after he got past the backstage security guards. (He had told the guards that he was "with the BOC"; BOC was nothing more than an abbreviation for their first startup: Books On Campus.) But in the entire summer batch of applications, the best story to tell about a successful hack was Ted Suzman's: when six years old, he had irrefutably disproved the existence of the Tooth Fairy. When a tooth fell out one day when he was alone, he put it under his pillow and then did not mention it to anyone. When he was with other family members, he made sure to keep his mouth closed until bedtime.

The trio has successfully hacked the world of Facebook apps; their app Graffiti is used by Facebook members to send illustrated messages to friends.[11] Technically, Graffiti is nothing more than a simple drawing tool and messaging app. "I love you," accompanied by a heart, is the message most commonly sent. It is free and is used by 1.5 million members monthly. Users have dispatched more than one hundred million drawings to their Facebook friends.

Mark Kantor is the sales guy for the Graffiti Labs team. He does not indulge in small talk; he has not created an online persona; he does not raise his voice; he does not smile a lot. He brings a steady seriousness to his work. And he is a very persuasive person.

The Graffiti app supports an extremely profitable business for the trio. Thanks to Kantor, corporate sponsors have been lined up, including BMW, Microsoft, Dell, Intel, Paramount, Comcast, HP, and Sharpie. Sponsorships have brought in about $600,000 a year. At the time that the founders applied to YC for the summer, Graffiti had already generated $2 million in revenue. This was the standout story among the startups that already had a business at the start of the summer batch.

Other than having Kantor minding the sponsor queue, Graffiti, the app, does not require much attention. Rather than endowing Graffiti with more features, the three want to embark upon a related, but new, venture, which is why they have come to YC. They are working on Graffiti World, which combines digital Legos and the social aspects of the building game Minecraft.[12] Graffiti World's users will be invited to submit drawings of objects, which other users will use to build scenes, their own Graffiti Worlds. Much coding needs to be done before it will be ready for a beta release, however.

Kantor and Tim Suzman live in an apartment in the Russian Hill neighborhood of San Francisco and their front room serves as the company's office. Ted Suzman has his own apartment in the Mission district, so the three are not housed as compactly as the Ridejoy trio. But to an observer watching them work, the Graffiti Lab's threesome is a seamless unit.

✦

In the middle of the summer, the Graffiti Labs founders find themselves, most unexpectedly, working on two different ideas simultaneously. The second idea began as a side project of Tim Suzman's—hackers cannot resist side projects. He wanted to do an improved version of "Ask Me Anything" found on Reddit. In Reddit's "I Am A" (IAmA) section, anyone can offer to field questions about their work or exploits or some aspect of their life that is unusual. The heading typically takes the form "I Am A—" followed by a noun, such as "Nightclub Doorman in Miami," and then the initials AMA, for Ask Me Anything. Questions, answers, and thousands of comments can quickly accumulate under a single entry. At Quora, another question-and-answer Web site, one question would attract answers from many people. Reddit's AMA is the converse: questions from many people are answered by a single person.

Suzman had a domain name, AnyAsq.com, where he could try something like Reddit's AMA. Where Reddit was cluttered and open to anyone to invite questions, however, AnyAsq would be visually clean and selective about who would field the questions. As a lark, he put up a couple of entries and got Mark Kantor and a few close friends to do the same. "I was a four-time MA scholastic chess champion. AMA," offered Suzman. One of Kantor's was, "I was an EMT in a major city at age seventeen. AMA."

Suzman mentioned it to a few YC batchmates, and Jason Shen of Ridejoy volunteered to field questions: "I was a nationally competitive gymnast for ten years. Ask me anything." Then Tikhon Bernstam of Parse also jumped in: "I cofounded Scribd.com." YC partner Harj Taggar saw Bernstam's tweet about it and, curious, paid a visit to the site. Not realizing that AnyAsq was connected to YC founders, Taggar liked what he saw and clicked to add his own AMA: "I'm a partner at Y Combinator. Ask me anything."[13]

The site got more contributors: a PhD candidate researching suicide who interviewed subjects within twenty-four hours of their attempting to end their lives; a member of Google+'s development team; an NFL player (a relative of Kantor's girlfriend). It is an encouraging start, and they do not know which idea they should pursue—the yet to be finished Graffiti World or the just launched AnyAsq. They consult Graham, who gives his blessing to their working on both ideas simultaneously.

Within a few weeks, the Suzman brothers and Mark Kantor have learned enough about the online advertising business to conclude that AnyAsq would have to attract tens of millions of unique visitors monthly if it were ever to generate significant revenue. In other words, it would have to become a widely recognized Web brand. They have put AnyAsq aside and are giving full attention to Graffiti World.

✦

It's the second week of August, and Demo Day looms. In the apartment/office, the three look ahead. Kantor recalls Graham saying at the very beginning of the summer, "What are you going to do to get revenue? We want to see curves." But they will not be able to show the graph that Graham is looking for. They have decided to proceed slowly, holding off releasing Graffiti World until they know more about which parts of their planned design will be utilized and which will not.

They will first test the Graffiti World prototype with small numbers of users and see what they do with the available objects. Traditionally, game developers like Electronic Arts would work on a new game for a year or two—"in the dark," Kantor says—and then release it, hoping for the best. He and his cofounders want to do something different: test a crude version on a tiny scale, get feedback, adjust, and try again. This would never be possible if users had to download and install software each time the program was tweaked, but because it will be a Web-based game played inside the browser, the code can be changed at any time without imposing any inconvenience upon the users.

All three founders are concerned about the competition for talent that will be set off once Demo Day is past and the summer batch's companies start to hire. Their recruiting should be helped, they figure, by being able to point to the success of Minecraft, the building game. Minecraft was created by only one person, Markus Persson of Sweden, and its graphics are not as sophisticated as Graffiti World's will be. Persson has made more than $50 million in a single year. Kantor remembers noticing the game just after it had been introduced, when it had only about one hundred users. He had written a note to himself: "Show Tim and Ted this. Looks kind of cool. Maybe we could buy it." He thought they could acquire it for about $5,000. Before he brought it to the attention of the Suzmans, however, Tim sent him a link to the game with a note: "Two hundred grand in the first week." So, no, there would be no acquisition for pocket change. But it did offer encouragement to creators of a new builder game.

✦

The NowSpots founders—Brad Flora, who is twenty-seven, and Kurt Mackey, who is thirty-one—lived in Chicago but did not know each other before applying to Y Combinator. A year earlier, Flora, the salesperson, had conceived the idea for a product, raised $180,000 from the Knight Foundation, and with a technical cofounder had got a rudimentary service going. With NowSpots' technology, newspapers can embed a box on their Web pages that is the size of a standard display ad on a Web page. Within it, regional and local businesses can inexpensively place advertising by directing to it the tweets and Facebook updates that they generate for their business's Twitter account and Facebook fan page. Flora's original hacker cofounder had left in March and, with the help of mutual friends, he had paired up with Mackey, the replacement hacker, just before the YC interview.

Flora is single. Mackey is married with four children: two girls, ages three and nine, and a pair of infant twins who have just arrived as foster children. For him, every day spent in California, away from the family, is acutely painful. It's mid-August, and while sitting at a table in YC's hall, he describes what has been most difficult. "I haven't struggled that much with the nine-year-old, 'cause I can get on the phone with her and she can express anything she wants to express and tell me she's sad or doing any of that stuff," he says. "I feel like she at least understands what's going on.

The three-year-old is hysterical because she does not. She gets on the phone and she's like, 'You come home now, Daddy.'" He laughs. "'No, not yet.' 'You come to my house?' is her favorite thing. So I try to video chat with her, just so she can see that I exist."

This time in California has been purchased at a dear price, but Mackey believes it has been worth it. "To me, *this* is the time. I'm never going to have the three months of completely focused, distraction-free time to build something again." The difference in age—and a difference in intensity—separates him from the youngest YC founders in the batch. "People go out and do stuff on Friday nights and I just work. This is it. I feel extremely—I don't know—'hungry,' I guess, is the word, and I haven't noticed that from everybody else. And a lot of it's because"—he laughs—"there's a lot riding on this for me compared to if I were twenty-two and had the Start Fund money and no kids to worry about."

Of Brad Flora, his cofounder, who is not present, Mackey says, "We clash a lot. But I don't think either of us would be willing to stop. We've had our dramatic times over the last couple months. Time seems to take care of them. It's not important anymore. Very much like being married." He laughs again.

Flora arrives and takes a seat with Mackey. Since July, NowSpots has been generating sufficient revenue that the two can pay themselves what Flora calls "normal person's salaries." Flora recalls how the two had turned down an investment offer from Excelerate Labs, based in Chicago and modeled after TechStars, in order to come out to YC.[14] Excelerate Labs called Flora "a traitor to Chicago," but now some of Chicago's angel investors who had declined to invest earlier are asking Flora if they can put money in.

Mackey had flown home to Chicago the week before to see his family. He decided to go at the last minute, so the airfare was eight hundred dollars. "I feel fine with Kurt spending that money because it keeps him in a good mood," says Flora. "We had a guy visit last week who is going to help with some sales stuff. He and I went to Target and he stood in front of two products, like, 'Oh, which one of these bars of soap do I want? This one's two dollars more than the other.' I'm like, 'Dude, you just wasted money.

Time is money right now. You just grab the first soap and put it in there and we'll go. We don't have time to be fussing with stuff like that.' I worry sometimes that that makes me too 'spendy,' but we're moving as quickly as we can to get stuff done."

Flora had gone back to Chicago himself three weeks earlier to deliver a sales pitch to the *Chicago Tribune*'s sales force. Looking over at Mackey, he says, "I've got to make sure we bring in revenue or else Kurt's going to leave me"—Mackey laughs—"for another one of these YC startups that is kicking butt."

16

FEARSOME

As Demo Day approaches, anxiety spreads among the founders whose startups are at an early stage, as they worry how they will be able to convince investors that their story is compelling. Their anxiety is not dispelled by Paul Graham and the YC partners, who gravely review and endlessly fuss over the material that will be presented. Of course, investors are going to impute significant informational value from the mere fact that the startups were selected as YC investments in the first place.

One founder among the summer batch's 160 is a university faculty member: Elizabeth Iorns of Science Exchange. She is a research assistant professor at the University of Miami's medical school. At YC, however, she is about to pass through one more graduation ritual.

It's two weeks before Demo Day, and Iorns and her two cofounders—her husband, Dan Knox, and Ryan Abbott—have been invited in for an inflection point meeting. Four YC partners—Paul Graham, Paul Buchheit, Garry Tan, and Aaron Iba—are waiting in the back room when the three come in. The last time the founders had seen the small room so full was when they had come for their interview as YC finalists four months earlier.

As everyone takes their seats, Knox remarks, "This is much less stressful than the first time we were in this room."

"Yeah," says Paul Graham. This completes the pleasantries. "So. You guys." His elbows are on the table and he massages his cheeks. "We wanted to talk to you because what you're doing is so different than the stuff we usually fund that it's not natural for us to understand you or give you the right advice. If you were writing an iPhone app, we'd know exactly what to tell you. But I'm worried that we might not have given you the right advice or something. Not that there's anything specific we told you that was wrong. I just felt we ought to talk to you more."

He turns to Iba, who had become a YC partner only a few weeks earlier. "Do you know what these guys are doing?"

"No, I don't, actually."

Graham faces the founders, "Tell him. You can practice. Pretend he's an investor."

Iorns begins. "So we're creating an online marketplace for science experiments." Her soft voice has a strong Kiwi accent.[1] "It's kind of like eBay for science experiments. At the moment there's a lot of outsourcing that's done internally within universities. And we're trying to create an efficient marketplace that allows people to conduct their experiments across all of the universities, which at the moment it's not possible to do."

Iba is impassive. Graham jumps in. "For example, if you need to do some kind of test that requires some expensive piece of machinery and your university doesn't have it, right? What you have to do is you have to get it done at some other place. So it already happens. But in a very disorganized way."

"It's actually a very, very large market," says Iorns, explaining that about $1 billion annually is spent "outsourcing" research—to other institutions, she means, not to other countries.

Knox picks up the thread. "That's U.S. only. We've modeled our business on other markets, like the IT market, where they have a proportion of their outsourcing going through a platform like oDesk or Elance. We're estimating approximately 10 percent of outsourcing will go through our platform." That would translate into revenues of about $100 million annually.

"How soon?" asks Graham.

"Next year," says Knox, deadpan. A moment later, he and Iorns break out laughing. No, $100 million cannot be reached that quickly, at least if the figure refers to their commission. The $100 million milestone in commissions would require $1 billion to $2 billion worth of research contracts passing through Science Exchange's system. Knox says, "We think the amount of outsourcing in academic science will trend from—"

Graham interrupts. "A lot of the reason I ask these questions, by the way, is to give you practice. *That* is what an investor will ask at that point. He's not going to say, 'bull-*shit*!' which is what most people would say. He's going to say, 'How soon?' which is investor talk for '*bullshit*.'"

"Is 'five years' generic enough? It's not going to be next year, obviously."

Iba suggests a better response to the investors' "How soon?" question: "Depends on how much money you give me." He laughs.

"That's a good answer!" says Iorns.

Knox adds, "'Cause part of what we need to do is get awareness of adoption of this platform."

"What do you need to spend money on?" asks Graham.

"Sales," says Iorns.

"You want to have a sales force?"

"Yeah."

"You want to do outreach in the universities. So educate people that this even exists."

Knox replies, "I think you might have suggested it but Sam Altman really drove home that having people on the ground is a good idea. So we've started—we've got four interns."

"And by that, are you talking about campus reps?"

"They're postdocs," says Iorns, referring to postdoctoral fellows who have completed their PhD and do research in a university lab for several years before looking for a teaching position.

"So don't call them interns," says Graham. "That's confusing. It sounds like they're working on your behalf. You're paying them a percentage—they're not your employees."

"We call them advocates," says Iorns.

"They get a stipend," adds Knox.

Graham frowns. "They shouldn't need it."

"Just to get them off the ground. It's small."

"They're doing, maximum, five hours a week at twenty dollars an hour," says Iorns.

Ryan Abbott, anticipating what Graham is about to suggest, heads it off. "It's impossible for us to make these campus tours where one of us would fly in and be at all these labs. It's actually impossible."

"It *might* pay, net," says Graham. "You should do the math! Would that pay, net?"

"Last two weeks, I was just traveling the whole two weeks," says Iorns.

"But if you pay a sales guy to go do that, how much would it cost? If you raise money from a VC and hire twenty salespeople and that was their whole job to go do that, they could do it," he insists. It's a suggestion that is similar to his advice to CampusCred at the beginning of the summer—*go hire salespeople!*—but university research labs would need a different kind of salesperson than those sent to call upon pizzerias and coffeehouses.

Iorns pushes back at Graham's suggestion to hire professional salespeople. "I'm not sure they would be as effective as a postdoc, who's already got the connections within that university." Science Exchange launched only a short time before and she says she hesitates to draw broad conclusions, but so far the company has had the most success when prospects already knew her from her own scientific work. "Almost 100 percent of them signed up and we had a really high posting rate." The same pattern could be seen among the postdocs: the principal investigators who knew the postdoc's scientific work trusted the person and were willing to post research projects on Science Exchange.

"How much do the postdocs make?" asks Graham. "The most successful postdoc. How much have they made so far?"

"Well, the most successful one has only been with us for a week. And she's made $6,000."

Graham beams. "That's pretty good for a postdoc!"

"No, she only gets 5 percent of that." The $6,000 refers to the value of the contract, not to the commission.

"She made three hundred bucks, though," says Graham.

Paul Buchheit wonders, "Do universities mind having postdocs also being sales reps?"

Iorns answers, "The good thing about postdocs—and this is why we deliberately chose postdocs—is they're not employees of the university. Secondly, they get paid hardly anything, so they're really looking for work."

"They're also highly mobile," says Graham. "They're going to be out of there, somewhere else, like Typhoid Mary—they're going to spread you to another university."

They are likely to have connections to three different academic institutions already, says Knox: the place where they hold the postdoc position, the university where they earned the PhD, and their undergraduate college or university.

Iorns reports, "I went to a couple of conferences last week—"

"How did those convert?" asks Graham.

"Very poorly."

"*In-ter-est-ing!*" says Graham. "Why do you think? Because we had great hopes about that, I remember."

"I know. What I found was that people were very interested in the idea. They seemed very enthusiastic. They engaged with me. They asked me lots of questions. They took brochures. They never signed up."

"This is why you need a sales dude who does nothing but hound. *You* get leads and *he* hounds them."

"I wondered if it's the difference between they're not actually at a computer at that time, so they don't sign up? 'Cause you're just giving them a pamphlet. Interestingly, the conference I was at last week was—it's only scientists funded by the Department of Defense, very elite, you have to be invited—no vendors, no advertising. So I was there to present my work. I had my brochures, so I kind of used it as a kind of marketing opportunity, in the sense that there was no one else there who was competing. But still. The conversion rate was bad."

"Maybe you've got to treat these guys as leads. Don't expect them to go home to their computers and sign up."

"In that case, I should have been more aggressive in getting their e-mail addresses. 'Cause I wasn't."

"You want to show them the idea and get their card and then have someone follow up," says Buchheit.

"Scientists don't have business cards," says Iorns.

"Just get their e-mail address," suggests Graham. "When you talk to them, say, 'Can I put you in the system?' "

"The enthusiasm made me think that wouldn't have been necessary," she says. "But clearly it was." She laughs at her own naïveté.

"They could have *been* genuinely enthusiastic," says Graham. "But it's just not their top priority coming back from the conference."

Iorns says that when she sends a personalized e-mail message about Science Exchange to a particular scientist, inviting the recipient to sign up, she gets a conversion rate of 30 percent. The YC partners declare this to be an impressive response.

Graham moves to a new topic. "If you had as much money as you wanted, how fast could you make this grow? Again, I'm asking this question because VCs are going to ask. 'What could you do with tons of money?' 'Cause they're in the business of finding things that, if you give them tons of money, will grow very fast."

"We have a list of all of the major research universities," says Knox. "We've ranked them by the amount of research funding they've got and the researchers." He wonders aloud whether Science Exchange should hire PhDs as sales reps and have them work their way through the list of the top one hundred institutions.

"You should do some back-of-the-envelope calculation," says Graham, referring to calculating the expected revenue growth if sales representatives were hired. "It doesn't have to be absolutely convincing. It just has to be plausible reasoning. When VCs ask you for estimates of how well you're going to do, they don't believe them. They want to see how you think. If you give a *plausible*-sounding story about how you can get to your billion dollars a year, five years from now, that's *very exciting* to a VC."

Knox begins to calculate out loud. "So, now, currently $50,000 worth of experiments through to—"

"Fifty thousand dollars over what period? Ever?" asks Graham.

"That's what we've done in the last three weeks."

"OK. So what is it per week, now?"

"It's kind of lumpy," Knox replies.

Iorns explains, "We had one experiment that was $10,000. The next one was $500."

"How many total experiments?"

"So far we've had eight that have completely gone through."

As a marketplace, Science Exchange needs to attract a balanced proportion of willing buyers and sellers—labs that need someone else to perform a certain experiment and labs willing to undertake the work. It is the buyers, however, who must show up first, posting on Science Exchange the details of the work that they would like performed. Then the staff at other institutions can post bids to do the work.

Science Exchange is not running an auction, however, and the work does not automatically go to the lowest bidder. The buyer can examine the reputation of the prospective suppliers and choose whomever he or she wishes—or decide not to offer the contract to anyone.

So far, twenty projects have been posted on Science Exchange's Web site. Aside from the eight that have been completed, two other projects have attracted bids but the buyers have not accepted any. In one case, seven bids were submitted for one project, but the principal investigator had simply gone quiet.

"She's just never been back to the site," says Abbott, "and won't respond to e-mails."

"Who knows? Maybe her kid broke her leg or something," wonders Graham, who thinks she may still respond.

Another possibility, says Iorns, is that the buyer has circumvented Science Exchange, making arrangements privately with one of the bidders so that her laboratory will not have to pay a commission to Science Exchange.

"Airbnb prevents this," explains Knox, "but at the moment we don't stop people from putting in their e-mail addresses and bid information. So she gets an e-mail saying you've got a bid on your project—here's the details, which includes an e-mail address—"

"Mmmm," says Graham, considering this possibility.

"So circumvention is possible and we're not doing any steps to prevent it. One of the big things that's different from Airbnb is we're all about real reputation and the person's identity. Because it's important."

"Has to be," agrees Graham.

"At the same time, it makes it easy for them to reach out—even if the e-mail wasn't there, you've got the name, the institution. One Google search finds their e-mail address, right?"

Graham predicts that the risk of circumvention hurting Science Exchange's business is going to be used by venture capitalists as an excuse for turning the startup down. "This is the thing: when VCs decide they don't like you, it's based on gut feel." He looks at Iorns as he describes what he thinks will happen. "'Cause you're a female founder. Or something like that, right? And they will use *that*"—the buyers' circumvention of Science Exchange—"as their reasoning. So beware. They reach for the easiest, most plausible objection—that you're going to get circumvented and that's why this isn't going to work."

"Well, circumvention is one of our least problems," says Iorns.

"No, *no*!" says Graham, excitedly. "I'm not saying it's a real problem. I'm saying, if people tell you this, don't think, 'Oh, I can talk him out of it.' It's what the VCs who don't like you are going to use. That's going to be it."

Iorns says she understands that mention of a marketplace immediately sets off alarms, but university-based science research has certain characteristics that make circumvention difficult. Universities will only directly pay approved entities and the approval process is extremely bureaucratic. For its commission, Science Exchange takes on the approval chores so the funds can be drawn directly from the university's accounts. If a researcher were to try to make arrangements privately with another university, she says, "you've just created a big world of pain for yourself because you probably have to pay on your personal credit card and get reimbursed. It's way more of a hassle than just using our platform. That's why circumvention is not really something that we've been very focused on."

"Your growth is limited by the people putting in bids?" asks Graham.

"No. The people posting the project—that's what's limiting us at the moment. We have an enormous number of people wanting to submit bids."

"You do? So you're the opposite of Airbnb. Airbnb is host-constrained. You're guest-constrained."

Eventually, the founders plan to expand beyond university-based research to industry. Most immediately, they need advice about whether they should undertake active fund-raising. Demo Day is two weeks away, but they've already met with a couple of venture capital firms and are receiving requests for meetings from others.

"If we were to aggressively go out there and try to get meetings—" Knox begins.

"It's stupid," Graham says. "Because Demo Day is designed for that purpose, right? You're invited to a banquet. And you're thinking, 'Well, I'll go get a hamburger on my way.'"

"Well, I just worry," says Iorns, "how competitive is it at Demo Day, in terms of trying to stand out?"

"*All the investors are there!*" says Graham. "Yeah, sure, there are sixty-three startups. But there's also like four hundred *investors* all in one place." If it were in the interests of the successful startups to drop what they were doing—working on code and meeting customers—and instead try to raise money on their own before Demo Day, then that is what they would do, reasons Graham. He notes that Science Exchange has already received additional investments from both the Start Fund and SV Angel, which puts them ahead of many of the startups in the batch.

Iorns looks up from her notebook. "The other note that I had was to come to you to get a good guess for our valuation," she says.

Graham looks down and speaks slowly. "The story about the market is fabulous. But it's so hard to predict for VCs because it's so weirdly different from other startups." He continues to speak in a slow, low voice. "It will all come to what they feel in their gut about you guys as founders. If you seem"—he pauses again—"you know, fearsome, like you're going to take over the world, then they'll think, 'OK, these guys are going to take over the world, and there's definitely a world here for them to take over,' right?"

He looks at Iorns. "You're a female founder. A lot of VCs, consciously or unconsciously, sort of automatically discriminate."

Abbott asks, "So we might as well ask for a large valuation?"

Graham doesn't answer immediately. Buchheit speaks up, "Might be better that way."

The valuation will be "constrained by ambition," adds Garry Tan.

Knox says, "I think we can compellingly make the case that there isn't much constraint on our ambition. If they push us on 'How big is this market?'—"

Graham breaks in. "It's not so much what you say, it's more like how you carry yourself. And there's not much I can advise you to do. I wish—I wish—" He stops, then starts again. "Do you remember all the guys at the fund-raising panel?" He's referring to an evening program back in June in which partners and YC alums had offered advice about fund-raising. Ryan Abbott had been there. Graham continues, "You probably remember Joseph Walla from HelloFax. He seemed like kind of a badass, right? All those guys up there seemed pretty badass. He seems badass now. But, man, during YC, he would have been below the median for badass-ness."

Iorns laughs, imagining what Graham and the partners had done to him. "He's been beaten into shape!" She laughs again.

"Remember the first day I said you have to undergo this sort of transition that produces James Bond? Remember that? *He has undergone that transition!* If you took the original Joseph Walla and, like, put him in the transition that makes people seem more confident, more resourceful, and tough—right?—that's what you get out."

Tan agrees. "The difference is all tone. Old Joseph Walla would be, like"—Tan changes his voice to sound indecisive—" 'Well, maybe, we think that there's something here.' Versus, by the end of it, he's, 'There *is* something here.' "

"It's all confidence," says Buchheit.

"Just take out all the weasel words," says Tan, and laughs. This is not so easily done in this case, however, since the chief executive of this startup is a research scientist. What Tan calls weasel words—*we-can't-be-certain*s, the *we-do-not-know*s and the like—are the qualifiers used by scientists to make what they say precisely accurate.

Buchheit continues in this vein. "You can't be, 'Well, if this goes right, if that goes right.' "

"We don't tell people stuff like this much," says Graham. "'Cause I'm not sure there's much you can do. Some people will become Joseph Walla."

"That's all I told Walla, though," recalls Tan. "I was like, 'Take out all the weasel words.' I tried to coach him that way."

"Really?" asks Graham. HelloFax had been in the winter 2011 batch.

"Yeah, when we were doing the Demo Day stuff."

Graham says he is not talking about what is said in a presentation, but about a change in the very way Walla carries himself all the time. "I've seen it *time* after *time*." He offers another example of transformation: "Justin Kan seems so terrifyingly fearsome now. And I remember he seemed—"

"I think it's coachable, though," says Tan.

Graham continues his story about Kan. "During the first YC batch, he seemed so clueless. He seemed like he just rolled out of bed." Kan must have undergone the transformation when on his own; Graham makes no claim of having played a part in the process and he declines to agree that fearsomeness is a characteristic that coaching can instill. He has just told the Science Exchange founders that this is a topic that he rarely brings up because he does *not* think it is something that the YC partners, or anyone else, can produce in others.

As a scientist, Elizabeth Iorns was trained to be careful in what she says and does. Her graduate education probably did not include a unit on how to project a Braveheart persona that will lead soldiers into battle. Ryan Abbott speaks up.

"The knowledge of the market—Elizabeth is *very* confident. Any question that anybody has, she's going to have *the* reason—"

"I don't think there's much you can do," says Graham, "this quickly, to change how much you seem like Joseph Walla. Really, the reason we got onto this subject is because you were asking what your valuation would be, and it's going to come down to their personal reactions to you. Which are going to be *impossible* to predict. Be fatalistic. Since it's due to this unpredictable thing, it's going to vary a lot between VCs. Realize that some VCs you're going to talk to are going to like you and some VCs aren't, and whatever reasons they give you are going to be bullshit. They're going to say, 'It's because people can circumvent you,' or they don't understand the market, or one of their standard 'It's not you, it's me' lines. Don't try and fight them. If they say, 'You're circumventable,' don't say, *'Oh, no! But we're not!'* Because what they really mean is, 'We don't like the way you look.' "

Knox says, "You said it, but Ben Horowitz, I think, also said it: 'Listen to the answer, don't listen to the reason.' So, no is no. Got it."

"So who knows?" Graham sums up. "Whether any will be interested, what the valuation will be—it's totally unpredictable. That's true in general, but most of all with you guys."

The session ends without a cheery note to neutralize the depressive speculation about investors turning Science Exchange away.

You don't receive accolades at YC for doing what you said you were going to do when you applied. In Science Exchange's case, it said it would launch a marketplace and it has done so. And it has managed to secure the customers who are the most difficult to find, the first ones who are willing to pay.

Science Exchange may not be able to present a face of martial fearsomeness to investors on Demo Day. But it remains to be seen whether all investors will slight Science Exchange for failing to appear like all the rest. Just as the startups in the batch have individuality, so too, the startups will discover, do investors, who do not always act as a herd.

17

PAY ATTENTION

"Watch what works!" Paul Graham instructs the founders. It's Rehearsal Day, a week before the Demo Day presentations. This begins seven days of nonstop rehearsing and critiquing by YC partners. This first day is directed by Professor Graham.

"You're going to get sixty-three data points about what slides you can read, what slides you can't, what things people say you can understand, who talks too fast. I was going to say who talks too slow, but it's impossible to talk too slow. You'll see—everyone's going to talk too fast."

Each startup will have two and a half minutes to present its pitch, just as it will on the actual Demo Day.[1] Jessica Livingston will call out "time" when the limit is reached. On Demo Day, she says, "I'm going to have a hook."

The order of presentations is randomly assigned, and each presenter is followed by a critique delivered by Graham, standing at the side or in the rear of the hall. Presentation, critique. Presentation, critique. His criticism is sharp, employing the directness that professors use with grad students.

The founders assigned to the first slot come up. One steps onto the small semicircular dais and one sits down to operate the laptop that controls the slides. When the speaker steps down, Graham begins. "All right, so you guys went first, so you guys made all the mistakes. You're like the front of the car, catching all the bugs. You've got to look at the audience and not at the screen when you talk. If you look at the audience, you force them to pay attention."

Graham has introduced the theme that will run throughout his critiques: presenters will have to work hard to hold the attention of the audience. On Demo Day, the sheer mass of presentations will make it difficult for an audience member to remain tuned in continuously. "Use the BlackBerry test," Graham advises. "Is my presentation the one where they're going to check their BlackBerry?"

The first presentation lasted only a minute and a half, a full minute short of its allotted time. "You should get to two minutes and thirty by saying the same words that much slower," Graham says.

The next presenter, like the first one, is a nonnative English speaker. Graham repeats a point made earlier. "If you've got a strong accent, you've got to speak slowly. I lost entire sentences of what you were saying. And *I* know you and have *some* practice with your accent. Most of the audience will be seeing you for the first time. If I'm losing entire sentences of what you are saying, that means you might as well have just not said them—you might as well have just stood there and moved your mouth, right?" When Graham says something more cutting than he perhaps intended, he softens its impact by immediately looking down at his notes and uttering a placatory "umm."

The next presentation is given by a founder from the UK, who begins in an affectless monotone and, worst of all, is speaking very rapidly, too.

"Wait," Graham interrupts after just a few seconds. "What did you just say? What was the beginning of that sentence? 'You're probably wondering'?"

The speaker repeats it: "You'reprobablywondering."

"You've got 'you're probably wondering' down to one syllable. Seriously." Graham repeats the phrase, speaking slowly, making every syllable distinct. "You're. Prob. Ab. Ly. Won. Der. Ing." The advice he has for this founder is offered to everyone: "Here's the trick for deciding how fast to talk. If it doesn't feel wrong, you're talking too fast."

Another speaker, this one U.S. born, comes up and apparently believes his presentation is easy to understand. It is not. He goes too fast. Graham instructs, "Say fewer words, slower." This will not be the last time today that he says this. In the teaching business, repetition and patience are necessary.

As Graham moves on to other points, he shows irritation at the way this speaker has boasted of the programming prowess of one of his hacker cofounders. The speaker has singled him out for learning to program when he was thirteen. Graham is not impressed. "Every hacker started programming at thirteen. It's actually rather late."

Every presenter, Graham says, must make very clear, at the beginning, what idea the startup is pursuing. "A lot of people will be sitting in the audience and will assume that the presentations are not of equal quality, that there are going to be some dogs in this batch. And they want to know what they're about to see a presentation for to decide"—Graham raises his voice a half octave higher—"*if they should even pay attention*."

Buzzwords and trite marketing phrases should be removed. To Graham, the generic phrase "We're going to disrupt this market" is the prose equivalent of stock photographs. "No one should use any slide that would fit in another startup's presentation," he says.

Tikhon Bernstam steps onto the dais and begins. "Hi, we're Parse. And we're building Heroku for mobile apps." A YC alum from a 2006 batch, he has already taken the Graham seminar in pitching investors and he does not speak too fast. In fact, he speaks extremely slowly, in an affectless voice.

Before he goes any further, Graham interrupts. "All right." The audience's laughter drowns out whatever Graham attempts to say—they can tell that Bernstam is going to be the first speaker to be criticized for speaking too slowly.

Bernstam is unaware of how he sounds. "What's that?" he asks Graham.

"Are you kidding? That was actually too slow." More laughter in the hall.

Bernstam begins again, speaking marginally faster but not by much. Graham allows him to finish.

"When I said you should speak slowly, I should have added: also, not in the intonation of a bedtime story or a hypnotist." Graham allows himself to tease the advanced grad students, the resident alumni, a bit more roughly than he would first-years. He resumes, "I expected you to say at some point, 'And then the lion said to the bear—' " This gets an appreciative laugh from the group. He raises his voice for emphasis. "You'll *put them to sleep*! Seriously. When you speak slowly, also speak emphatically!" Speaking emphatically comes naturally to Graham.

Neither the founders nor Graham's YC partners have his ability to pithily summarize a startup's idea. After another presentation, he says, "This is another one where I feel people won't even get the revolutionary nature of what you're doing. It's almost like your presentation conceals what you're doing." He exclaims in a high-pitched voice: "It's in there somewhere!" He then renders the idea crisply.

Michael Dwan of Snapjoy, the photo storage site, presents, offering the tagline "Organizing the world's memories." When Graham begins his critique, he asks for a less abstract tagline. "Seriously," he says. "People will see that and they'll think, 'All right, whatever. I'll just check my e-mail.' They're not going to do the extra work to figure out, 'What does he mean by organizing the world's memories? Oh, I understand! He means collect their photographs and then expand into other file formats.' What you should say is, 'We're going to host everybody's photos.' "

He is not done with Dwan. "You've missed the *biggest* thing you then have to have in a presentation like that, which is, why *you*? Everybody knows it's going to be a good thing to host everybody's photos. Everybody's got photos. You'll have millions of users. They'll pay to have their photos hosted. That's huge. But that's a giant prize. Why are *you* going to win it? Why isn't there something else that's already the right answer?"

Graham does not abandon Dwan to despair, however. As Graham says in the YC User's Manual, "We have a lot of practice cooking up take-over-the-world plans." He draws on the example of YC's all-time most successful company: Dropbox. He tells Dwan, "You've got to say, this is one of those things, like Dropbox or search engines, where there are a ton of people already doing it. You say, 'How are we going to win? There's already so many things hosting photos. Well, the sign that there's room to win is the fact that there's so many! That means nobody's got it right.'"

Graham continues. "Then you have to make the claim about why you've gotten it right. You've got to get into that. It's really good if you can say, 'All right, I know what you're thinking—why us? We're so arrogant. We're like these smarty-pants programmers, we think we can write something, host people's photos online and display them with some nice JavaScript, and we're somehow going to get all the users and not Apple?'" He chuckles. "Explain why *you*."

Up next come Matt Holden and Sean Lynch, who, when the group had last seen them at Prototype Day two months earlier, were working on Splitterbug, the expense-sharing app for smartphones. It had not seemed like an idea that could grow into a sizable business, and this thought had concerned the two founders, too. They had decided to throw Splitterbug out and start anew. Holden introduces TapEngage, an ad network for serving interactive ads on iPads and other tablets. A former program manager at Google, he speaks conversationally.

"That was a good presentation!" Graham exclaims afterward. "That's what a good presentation tone feels like." He faces the founders in the hall. "Did you see how it was sort of conversational and moved along? He didn't go into super-much detail but he explained what they're doing. That's all the investors need: 'I understand what they're doing. It seems like it could be a big deal.'"

Jason Shen of Ridejoy follows. His startup's tagline is "We're the Airbnb for rides." He has some impressive numbers to share. "Each year, Americans spend nearly a *trillion* dollars on transportation—just consumer spending alone. And last year, we drove over *three trillion* miles. We think this is a really exciting opportunity and we can't wait to show you more."

When writing on his personal blog, "The Art of Ass-Kicking," Shen projects fearlessness. But today, standing before the assembled group, he seems ill at ease and relies too heavily on reading his prepared text word for word. The only time he speaks in a natural voice is when he talks about the shortcomings of Craigslist as a ridesharing service. Graham notices this. In his critique, he says, "You became strangely animated when you talked about how Craigslist sucked." The audience laughs. Graham wants everyone to draw a lesson from this. "You should *choose* the part of your presentation where you want to be most animated. You don't want to be animated the whole way through—you'll seem like a used-car salesman. And you don't want to be like you're telling a bedtime story the whole way through, either. You want to be animated at some point, and you want to *choose* the point when it's animated." If Shen gets animated when he talks about Craigslist, "the audience will assume you have done that and they'll be confused because—" He pretends to be an investor in the audience on Demo Day who thinks to himself, "They're talking about how bad Craigslist is and this is clearly the most important thing about their startup. I don't understand."

Graham will not let the "trillion dollars" spent on transportation go unremarked upon. "Don't include macroeconomic stats like that. That just washes over people," he says. Any mention of "trillion" will bring to mind politics, not business. "Don't even put that in. If you're going to make some kind of estimate, do it from the bottom up. This is like saying, 'We're a software company. Software is a subset of business. Business last year generated—' " The audience laughs again.

Dropbox comes up several times in Graham's critiques. He has used Dropbox as a source of inspiration for Snapjoy as it tries to make its way amid many photo-sharing competitors. He brings Dropbox up again after Kicksend's Pradeep Elankumaran has spoken. Kicksend's users can easily send large files to someone else, a service that does not, on its face, compete directly with Dropbox, and Elankumaran never mentions Dropbox. But he has said that "file sharing is broken." That, Graham says, "is not going to be very convincing if the Valley is buzzing about how Dropbox just raised money at a valuation of $4 or $5 billion. You've got to be clearer about why people can't just use Dropbox. You can't just sort of hand-wave and say, 'Oh, it's got complex permissioning.' Obviously, a lot of people can use it. So you've *got* to say why you're better. Everybody is going to say, 'These guys are competing with Dropbox. Dropbox is kicking ass! That means it's going to kick these guys' ass.' That's what these guys are going to be thinking by default. So you've got to make most of your presentation addressing that."

Even if each presentation fits tidily within two and a half minutes, the procession of disparate ideas, without thematic grouping, will be stupefying. Today, after a numbing twenty-one presentations, a break is called. Two-thirds of the batch still wait for their turns.

In the first YC batch in summer 2005, the eight startups could make their individual pitches on Demo Day without becoming part of a blur in the investors' minds—each had fifteen minutes for its presentation (and only about fifteen investors, Paul Graham's friends, were present to hear them). As the roster of presenting companies grew, the time allotted to each company had to be shortened. Investor interest in YC grew apace, to the point where investors could not all fit into the hall, so Demo Day became Demo Days. Next week, the sixty-three startups will give their individual presentations to one group of investors on Tuesday and then again to another group of investors on Wednesday.

The YC User's Manual tells founders that even in the very best case, when an investor is paying close attention, information equivalent to only four or five sentences will be absorbed from a presentation seen for the first time. One of the sentences should be a genuine insight, something that will surprise—"If a statement is not surprising, it's probably not an insight," the manual says. Investors may remember far less of the presentation than this, however. A single noun may be all that sticks in the mind afterward. "Something to do with chat." "Something to do with databases." The manual says about a third of the startups on Demo Day will do no better than this.

A single noun? As a means of conveying factual information about the startups, a marathon session of presentations is perfectly awful. The persistence of the format reflects the seemingly indestructible desire of investors to meet founders in the flesh and gather impressions that cannot be conveyed by Web pages or e-mail or videos or phone or webcam or any other means. YC insists on meeting founders in person, albeit briefly, before making its own investment decisions. For the angel investors and venture capitalists who will come next week, a live glimpse of the founders, in the same room, is sufficiently valuable that they will come and subject themselves to a long, long day of presentations. If their attention flags, they will put their heads down and work on e-mail—as Graham warns the founders incessantly during rehearsals.

The investors will not come to Demo Day with checkbooks open and pens uncapped. If a presentation interests them—and a single noun and a favorable impression of the presenter may be all that is necessary—then the investor will seek out the founders of that startup in the swirl of conversations that will come at the end of Demo Day and have the first of many chats that will precede the decision to invest. The User's Manual explains that on Demo Day investors will have a printed list of the startups in their hands. As the program proceeds, they will circle the names of startups that they would like to speak with afterward. It advises, "The goal of your presentation should be to make them circle your name—to make them say to themselves, 'This could be big.'"

✦

The rehearsal resumes as the middle group of twenty-one companies begins. The presenter of the first one speaks in a flat monotone. With this speaker and some others, whose presentation skills could be described as falling in the bottom decile of the summer batch, Graham is sparing in his criticism. He pushes hard only on those he sees as able to fix the problems he identifies. For the others, he is candid—he can't help himself about that—but shows a gentle manner that he does not use for other founders.

"You don't seem very excited about your product," he says to one presenter. "I know that's the way you are. To the audience, it will just read like you don't think it's very good. I don't know what to do about it. I wouldn't completely change how you seem but maybe be a bit more excited."

Another presenter, who not only speaks without emotion but also has a soft voice that is difficult to hear, draws from Graham a suggestion that he memorize his presentation. "You're a very soft-spoken guy, not like a big carnival barker," he says. "It might be better to write it out and then you'll feel at ease just delivering it from memory." The suggestion is not directly matched to the problems that are manifest, but at least it gives Graham something to suggest.

Tom Blomfield of GoCardless comes up. "We are replacing credit cards online," he opens, and describes the startup's work on creating a low-cost payment system for online merchants that does not use existing credit card networks.

"That was pretty good," says Graham. "This is a good example of how the goal on Demo Day is to make people feel like they can't afford *not* to come find you afterward." Investors will think, "This sounds big. I don't know if this is real. But it might be real. And I'm going to be in big trouble with my firm if I miss this deal and it turns out to be one that goes public."

A presenter shows a graph depicting projected revenue. This gives Graham heartburn. "Don't show projected revenue," he says. "If you don't have any revenues yet, don't lie. You don't *have* to have a hockey stick. Only show it if you have one. Don't make one up."

✦

Forty-two companies down, twenty-one more to go. After another break, the presentations resume. Chris Steiner, of Aisle50, is a standout. Introducing the idea—"Groupon for groceries"—he speaks clearly, expressively, slowly, and confidently.

"That was a great presentation!" Graham says at the end. He turns to the audience. "Doesn't that sound compelling?" Then he turns back to Steiner. "You sound commanding. You sound like you know this business. You sound like an industry expert." This is high praise to receive as a newcomer to the grocery business—before coming to YC, Steiner had been a writer with *Forbes*.[2] Graham just cannot say enough about this presentation. "That was great!"

Near the very end, Hiptic's turn comes. George Deglin speaks while his cofounder, Long Vo, sits at the laptop, changing the slides. Hiptic plans to offer a "stylish single-page Web site for everyone." The service has not yet launched, but Deglin has a slide to show what a personal page on Hiptic would look like. "Let me start by telling you a story," he says. "This is my cofounder's girlfriend, Linda Le." The wall fills with the image of an extremely attractive young woman whose picture is composed in the unnatural way favored by a fashion photographer. In fact, Le *is* a professional model. The audience takes this in immediately and somewhere in the group an appreciative whistle rings out above the commotion caused by a presentation that is unlike all the others. Deglin continues reading his script. "She is actually one of the most famous cosplayers in the world," referring to a subculture whose name is derived from "costume play." Le is known to cosplay fans as "Vampy." Deglin says she is the subject of many online profile pages, some of which she has set up, others set up by fans. "What if she could take all of her latest content that she wants to share with all her fans, her best YouTube video that she just produced, her latest photos, and condense them down to a simple stream?" He points to the dummy mock-up for her on Hiptic. "What about something like this? This is the kind of profile that I think all of us want to have on the Internet. It represents you." Many voices in the audience fill the hall when he is done, as founders remark to their neighbors on Le's profile.

"All right," says Graham. "This is a problem that sometimes arises. If people use a demo involving beautiful women, no one will pay attention to what you're saying about this startup." The audience is relieved to have this acknowledged and laughs. Graham turns to the audience. "Do you remember anything he was saying?" More laughter. "The audience is mostly men! This has happened before." He turns back to Deglin and Vo and suggests that they use a man's profile in their presentation instead of Le's. Noises of protest come from the audience. He raises his voice to be heard above the laughter. "Or no one will pay attention to what you're saying. I'm serious!"

In every batch, about one-third of the startups decide to change their original idea, like the teams that have settled upon TapEngage and Ridejoy. Another team, Zach Sims and Ryan Bubinski, began with the idea for establishing a reputation system for programmers and then switched to BizPress, Web sites for small businesses. They now present their *new* new idea, the one that has replaced the replacement: Codecademy, offering to teach anyone how to program. It is a service designed for someone like Sims, the nontechnical founder and political science major. Whether it will appeal to others, however, is untested. The late decision to change course has left insufficient time to get even a rudimentary site launched yet.

Sims, who gives the presentation, can only point to indirect evidence that there will be demand for a service such as Codecademy's. He has a potpourri of facts testifying to general interest in programming: 5.5 million computer science books sold every year and a 50 percent enrollment increase in first-year computer science classes at universities across the country. Online education overall is growing 21 percent year over year, he says. Bubinski has taught undergraduates JavaScript, Ruby, and other languages; as for himself, Sims says vaguely that he "has spoken on educational technology" in college classes. Sims boldly claims in his conclusion: "We're the team to help bring programming to the masses."

This draft version of the presentation is not particularly compelling. It is really nothing more than an enumeration of some signs that the world out there seems interested in programming. Details about what Codecademy is actually going to do are absent, as is any description of the business. Graham says, "I think you should talk a little bit more about why it's a big deal. Otherwise, investors will just think, 'Oh, these guys are working on some open-source project. It's never going to be a big company.'"

Considering the time lost before the founders came up with the idea for Codecademy, they cannot be faulted for having little to present at this late date. But they have not grabbed the room's attention, either. When Graham finishes critiquing, the next company gets its turn and Codecademy's presentation is quickly forgotten, with nothing memorable to distinguish it from the pack.

18

GROWTH

TechCrunch has been crowded with announcements of YC-funded companies that are launching. Interview Street on August 6. Both Leaky and Snapjoy on August 8. Stypi on August 9. Envolve on August 10. Both MobileWorks and Picplum on August 12.[1] This is a new problem that has appeared since the large expansion of YC with this summer batch. Too many YC companies are attempting to formally launch their products at the same time, making it difficult to draw attention to any one. The "YC-funded" mention in a TechCrunch headline no longer stands out.

In this respect, Zach Sims and Ryan Bubinski are fortunate that they got such a late start on Codecademy. If their site were ready to be launched, it would be stuck in the "YC-funded" crowd trying to get attention on TechCrunch. It's Thursday, August 18, the day after Rehearsal Day, and their site is not close to being ready. They have cobbled together only a single course, an introduction to JavaScript with eight micro-lessons. It's painfully rudimentary and might fall short of even a low bar for minimum viable product. Before releasing it, however, they need to get some feedback without launching. Hacker News seems like a good place to quietly invite hackers to pay a visit to the site and critique what Codecademy has so far.

Midmorning, Sims posts a notice on Hacker News, "Show HN: Code cademy.com—The Easiest Way to Learn to Code," and he and Bubinski head out to get bagels for lunch.[2] In the car, they decide it would be a wonderful thing if they manage to get fifty concurrent users on the site. Sims has a mobile app on his phone that shows the number. As they drive, they laugh as they see that they may hit their target: thirty concurrent users, forty, and then more than fifty. As they stand in line to get their food, they check again: now the number is in the hundreds. And the announcement is spreading fast through the Twittersphere—hundreds of tweets have appeared in a blink. It has also been up-voted to the front page of Hacker News, which means it is going to get a lot more attention, bringing more visitors to the site. Realizing that their site might crash under the load, they head home posthaste.

At their apartment, they look at the monitoring app on one of their computers, which displays traffic with an arrow on a speedometer. They have more than a thousand concurrent users and the arrow sits inert at the farthest edge: it no longer functions.

Reddit picks up the Codecademy story.[3] The first person to comment complains that "there is only one lesson." After another person says, no, there are eight lessons, the first commentator responds, "OK, *maybe* a loose eight. But it was all incredibly easy and didn't introduce much of anything." Another commentator, who was "absolutely terrible with code," is more positive: "After the first two lessons, you know what? This shit ain't half bad."

TechCrunch notices the attention and contacts the two founders for a story it is preparing. The founders, leery of being another "YC-founded X" story, do not mention that Codecademy is part of YC. The resulting TechCrunch story could not have been more positive: "Codecademy's initial signup process is very clever: there isn't one, at least at first. As soon as you land on Codecademy.com you'll be prompted to complete the first lesson." It acknowledges that the few lessons do not go very far, but Codecademy "clearly has loads of potential for one key reason: it actually feels fun."[4]

The attention on Hacker News, Reddit, and TechCrunch produces more attention and more visitors, and the site does go down, briefly. Bubinski quickly puts up a sign-up form on the home page, inviting visitors to submit their e-mail addresses while Codecademy is readied to reopen. When the site is brought online again, only twelve minutes have passed—yet more than ten thousand e-mail addresses have accumulated.

The site uses Heroku and MongoHQ, so the founders send out emergency pleas for help to James Lindenbaum, a Heroku cofounder, and to the MongoHQs, their summer batchmates, and receive emergency aid, including help rewriting their databases. The Codecademy founders end up going seventy-two hours without sleep before collapsing.

✦

On Sunday, Demo Day is just ahead. Founders are in YC's main hall working on presentations, taking turns on the dais to rehearse. One YC partner sits in a chair at the back critiquing. Other teams huddle over laptops, waiting their turn. They use the time to review their slides and obsess about imperceptibly tiny tweaks.

The Rap Geniuses—Tom Lehman, Ilan Zechory, and Mahbod Moghadam—head into the meeting room in the corner for office hours with Sam Altman. They had started the summer in an enviable position, with a Web site growing at a healthy clip, seemingly on its own. The prodding from Harj Taggar to expand beyond rap has not produced action. Rap Genius remains what it was at the beginning of the summer, a site branded for rap lyrics. Yet the site's growth curve is steeper than ever. Altman is impressed and has been considering making a personal investment in the company as an angel. He has held off making a final decision until he hears back from an attorney who specializes in intellectual property issues. Altman wants an expert's opinion on the legality of Rap Genius's posting of copyrighted lyrics.

When the Rap Geniuses come into the office, they bring out a graph showing the number of unique visitors monthly. With a glance, one can see an acceleration in the growth. This will be featured in their Demo Day presentation.

Altman speaks rapidly, as he always does: "When you show this graph, make sure to make the point: 'All right, look—we're going to show you a user graph. Most YC companies get up here and they show you cumulative registered users. Like fuck that! We're going to show you active.' That's unusual. You're counting this the fair way."

He suggests that the team run through the presentation right now. "Go."

Lehman has rehearsed so many times that he can adopt a presenter's voice instantly. "OK, guys, we're Rap Genius. Before we start our presentation today, I have a little quiz for you guys." The test question has changed since Prototype Day. This time, the lyrics refer to Malcolm X—and to jeans. "This line sounds kind of cool, but what is Kanye West talking about here, what does this line actually mean?" He does not wait for an answer. "So this is the problem that Rap Genius is trying to solve. Rap Genius is a Web site where you can read explanations of confusing and interesting rap lyrics. So this is a demo right here of our live site and, as you can see, you can click any orange lyrics for explanations. This is the 'Good Morning' page on our site where you can read the explanation of that Kanye line. Now, we have about 250,000 explanations on the site, tens of thousands of songs. Who writes the explanations? Well, any member of our community can write an explanation. All the explanations are crowdsourced."

He reviews the site's traffic growth and shows the graph with the steep upward curve. "This isn't some promising results from an early startup that just launched on TechCrunch. This is a huge site. We get three million unique visitors a month. We've been out over a year."

In addition to covering English-language lyrics, fans are posting explanations for rap lyrics in other languages, too, such as French and German, he says. Fans are moving beyond music, too. "We're getting poetry explained on our site by our community, we've got the Bible up there—Moses, Genesis chapter one lyrics. We've got law—Bill of Rights on there. This is all of text."

Lehman swings into his conclusion. "What have we learned today? Right now, it's game over for all of the crappy lyrics sites—we're taking over the lyrics game. Done. Tomorrow, soon, it's going to be a social annotation platform for all of text. And of course, explosive growth."

Without a moment's pause, Altman begins his critique: "It was OK. You underserved yourself. The presentation was not as good as the company is. A couple things. I like the thing at the start and I think you present really well. Beyond that, you talk about the problem is that people want to know what lyrics mean. But that's not really the problem. The problem is that lyrics sites are terrible."

Lehman shouts out, "I forgot to mention the 2 percent!" This refers to his team's discovery of a source that says 2 percent of Google searches pertain to lyrics. Two percent of a gazillion is a colossal number, so it prompts disbelief. When Paul Graham heard it mentioned during earlier office hours, he thought he had misheard: how could lyrics draw so much interest?

"Yeah! You need a slide of that," says Altman. "You can't take any risk of forgetting that. People should see that. The way I think about structuring this generally is—problem. The problem that is real is that lyrics sites are *terrible*. Say, 'Look, lyrics sites are terrible. They're crazy flash—they try to make it on ringtones. *And* they don't even give users the number one thing they want, which is explanations. So we built a product that does. But before we show you that, we're going to show you our growth, because it's really working quite impressively.' Go to the growth graph. On that graph, what's the compound weekly growth rate?"

"We grew 50 percent this past month."

"Maybe say that. Fourteen percent week over week—that's killer." Altman says that other presenters will talk about 40 percent monthly growth rates, but their base is tiny. "No one has absolutes anywhere near you guys. This is a little aggressive, but I was thinking about—just to try to put in perspective how big you guys are—I bet if you total up the monthly uniques for the other sixty-two companies—"

Ilan Zechory sees Altman's point before it is complete and laughs. "That's so aggressive."

"You guys—part of your charm is you guys are aggressive, and if you get up there and say, 'We have 2.5 million uniques. If you total up the monthly uniques of the other sixty-two companies you're seeing today, they have less than 2.5 million uniques. *And* we're not even going to charge you guys sixty-two times the pre-money valuation'—that'd be kind of funny."

Lehman says that the company did something similar earlier in the summer, claiming in a job posting that their startup was "the fastest-growing summer startup," but "PG got a little upset at us—about the idea of us dissing the other startups, or whatever."

"You guys, you've got to talk about the size of the market. Two percent of Google searches are for lyrics, which is thirty billion per year. *And* a little bit about how you're going to monetize. The amount of money spent in the music industry in general. *Some* financial metric of how these other sites are monetizing, or like how much money is spent total on music-related things. Investors are going to believe you that it's a huge market, that there's a huge number of users. But they need to know that there's actually a lot of money they have a reasonable chance of putting their hands on as well."

For the long-term vision, Altman suggests that Lehman say, "Look, we started this to explain rap music. 'Cause it's something we wanted for ourselves. We had a suspicion a lot of other people do. But organizationally we're seeing this user behavior where users are trying to use Rap Genius to explain a lot of other text. We've seen them use this for the Bible, whatever, whatever. We think we have the opportunity to take these same principles, growth via search and annotation of text, and make something that works at an incredibly broad scale."

He warns them about appearing unfocused. "You don't want to do a presentation about rap and then talk about the Bible. You want to say, 'We built this incredible thing, it's growing like wildfire, and we're observing this user behavior, users wanting to use this for other things.'" He asks, "Can I try to give the pitch?"

All three founders say at once, "Sure!"

Altman begins, using the Kanye West lyrics, then stops. "You know what? This doesn't work. I'm going to skip it." He begins again.

"We are Rap Genius and we are a lyrics site that is actually really good." He uses the slides on Lehman's laptop as an approximate guide, explaining why lyrics sites are awful and why Rap Genius's is great, and describing the market's size and the company's traction. He stops again and observes that this has taken only a minute.

He resumes: "We're happy with this growth graph. But we really want to see how this gets to fifty million. And we think we're on a path there because we've identified this behavior in our users. Users aren't just using this for rap lyrics. They want this for all other kinds of music. And not just that. We've seen some crazy shit." He clicks the next slide. "Here's the Beatles. That kind of makes sense." Another slide. "But here's the Bible. And here's T. S. Eliot"—the slide shows "The Love Song of J. Alfred Prufrock." Before clicking to the next one, Altman mentions, sotto voce, "That's my favorite poem, by the way, of all time. I think I can sort of recite that from memory." He switches back to his presenter's voice. "We've seen this behavior where this thing that we're doing, this ability to find information about text, or find text that a user cares about in the sense of community, really appeals much more broadly than we ever thought. So very soon we're going to be expanding this to other genres of music."

Altman stops to ask the team, "Are you guys already live with rock?"

"Not yet."

"So you should say, 'We just started playing around with rock. We'll have a live separate site soon. But this is really just the beginning of what we can do here. And we're going to win because we understand how to do this. We understand the SEO' "—search engine optimization, or the tricks that increase the chance that one's Web site will show up at the top of search results. " 'We've already got a super-vibrant community. At this point we can just turn on all the genres really easily. So what do we want to leave you with? We're going to win in lyrics. These things build up a momentum of their own. You have a graph like this, it just keeps going. We're going to get 2 percent of Google searches. Tomorrow, we're going to expand this and we're going to take this behavior we've seen into these other things. And we have crazy growth.' "

He has reached the last slide. Something is missing, he realizes. "Why do you not have something in here—you had at some point—tweets about how much your users loved you?"

Zechory explains, "PG was like"—he pretends to yell with a voice filled with disgust—" '*Never put a tweet in a presentation!*' "

Lehman also performs an impression of Graham: " '*Bush league! Tweets are for losers!*' "

"The only reason I ask is, lyrics sites are generally something so hated," says Altman. "So maybe you just talk about it on your growth slide. Say, 'Not only are we growing like crazy, our users actually love us, too. We have a passionate user base.'"

"So do you think we should kill the opening quiz?"

"In general, the first fifteen seconds of your presentation are the worst time for a gimmick. Because the first fifteen seconds of your presentation are when investors decide to pay attention or not. If I were you, the first things I want to say in the first fifteen seconds of my presentation are: we built a great lyrics site and we have incredible hockey stick growth."

Mahbod Moghadam says, "The hard part is, when we say that, what we've noticed before is, as soon as they hear 'lyrics site,' they're done. So we don't even want to call it a lyrics site. What I think it really is, is a musical Facebook. What MySpace was supposed to be and then it got ruined."

"Yeah, you don't have to say 'lyrics.' You can talk about 'a community site for music.'"

Altman leaves the room for a moment to ask Graham something and the founders talk about plans to launch StereoIQ for rock lyrics and CountryBrains for country music. They had come close to buying Genie.us to serve as the overarching brand, but Emmett Shear had insisted, "Listen, you guys, whatever you do, just get a .com. You need to have a .com."

Altman returns. "PG thinks it would be more impressive if you guys said something like, 'We have more monthly uniques than Foursquare.'"

Lehman asks, "What about saying we're bigger than the other sixty-two combined?"

"I don't want to say that," says Zechory. "I would veto that."

"Why don't you want to say that?"

"Because it's dick. There's a lot of our friends out there trying to raise money."

Altman intervenes. "The point I'm trying to make is 2.5 million uniques is unusual. If you go through the history of YC—you know what? Maybe you could say—"

"That we might be the biggest site in the history of YC at Demo Day," says Lehman, completing the sentence.

Altman repeats the crucial qualifier—"*at Demo Day*"—and says, "I'm almost sure you can say that."

The founders want to talk about fund-raising: what are their chances of doing a Series A round with venture capital investors? Altman says he'll answer their questions, but he reminds them that he is a prospective investor himself and cannot dispense objective advice. "You *really* don't want to ask someone negotiating against you in a fund-raising environment for fund-raising advice," he says, and the Rap Geniuses laugh. "All right, ask the question again."

They ask about Series A rounds and valuations, uncapped notes and capped notes, the topics that all of the founders in the hall are talking about at tedious length. Altman urges them to consider setting a valuation and doing an equity round now, rather than a convertible note that will convert into equity at a valuation to be set in the future. "So if the economy does fall apart," he says, "the price is locked in."

✦

The MongoHQs, Jason McCay and Ben Wyrosdick, are out in the hall. Word about their startup has been circulating among angel investors and venture capitalists in advance of Demo Day. They have been spending recent days in a blurry succession of meetings with prospective investors, including some of the Valley's leading venture capitalists on Sand Hill Road.

The informality of YC's own screening process when interviewing finalists and YC's hacker culture, valuing technical skill above all else, does not prepare founders for the starchy culture of Sand Hill Road. Before his meetings, McCay had thought the venture capitalists would look past a rough presentation and lack of polish, and now he sees he was wrong. Appearances clearly do matter.

So too does preparation. "You've got to be focused," McCay says. The questions in these meetings had come fast and furiously. Before he was ready, he had been asked, "You're going to raise $5 million. Tell me your first ten hires." He explains, "And if you said, 'Three engineers,' they don't say, 'OK, three engineers.' They say, 'What exactly are they going to do?' They ask very specific questions. When you think about it, it's not unreasonable at all. But I went in there not being prepared to answer the questions."

Back in the middle of the summer, the two MongoHQ cofounders had thought that they and their wives might all agree to stay in Silicon Valley and not return to Birmingham. But the four are having second thoughts about staying, mainly because of pressure being applied by their parents, who, as grandparents, want the families to return. Wyrosdick describes what he and McCay are hearing: "When are you coming home? Don't you believe you're taking our babies away at their most precious time of their lives? We're never going to get these years back." The two cofounders feel pressed enough by business-related challenges; additional pressure applied by their extended families is almost unbearable.

Five years earlier, in YC's early days, Paul Graham had told graduating seniors at MIT that recent grads can live far more cheaply than engineers who are in their thirties. "The guys with kids and mortgages are at a real disadvantage," he had written then.[5] That was in a different era, when seed capital was scarce and founders would have to live on ramen. The MongoHQs had the financial means to handle a permanent move to Silicon Valley, notwithstanding "kids and mortgages," that they had before the start of YC. But as their case shows, nonfinancial complications can hamper mobility, too.

✦

The YC partner who over the years has been the most visible angel investor is Paul Buchheit. If Paul Graham is the professor who teaches his own fully elaborated General Theory of Startups, the other Paul is not professorial at all. He eschews grand theories and favors empiricism: what works, works.[6] In his own professional life, he has followed whatever path seemed most likely to lead to interesting work.

After college, as a bored engineer at Intel, Buchheit decided to look for something that was more engaging than what he was doing. The idea of joining a startup that would likely soon die was just fine—he was single at the time and did not have a mortgage to worry about.

In June 1999, he applied to Google, which then had only twenty-two employees. He was fairly certain that AltaVista would soon destroy the little startup, but he could see that Google had smart people and offered work that would let him learn something, and that's all he was looking for. He told Jessica Livingston many years later, in *Founders at Work*, "It worked out well, but it wasn't like I saw this company and said, 'Oh, wow, this is going to succeed!' I just thought it would be fun."[7]

He made his first angel investments after he left Google in 2006, and continued after cofounding a new startup, FriendFeed, the next year.[8] In 2010, he wrote, "I started investing in startups with the assumption that I don't know what I'm doing (which is always true), but that the only way to actually learn anything is through experience." His goals were to invest in a variety of companies, learn from and help those he could, and "hopefully not lose too much money while doing so."[9]

Angel investors assemble a portfolio, making lots of small investments rather than a few large ones. Their emotional tie to any given startup is limited. Buchheit's experience is instructive. After two years, he had invested $1.21 million in thirty-two companies, an average of only about $38,000 in each. By 2011, those among the thirty-two that had been acquired had brought a return of $1.34 million, or about a 10 percent gain over the two years. His rate of return could still go up: about half the companies were still alive and yet to be counted because they had not yet had an exit.

Buchheit's two best investments, each of which yielded more than a tenfold return, were Heroku, the YC startup acquired by Salesforce.com, and Mint, which had been acquired by Intuit. "Unfortunately," he wrote, "they were also two of the smaller investments, proving that I don't know what I'm doing, or at least showing that I need to make a point of investing more money into the best companies."[10]

Having begun investing in YC companies almost from YC's start, Buchheit pointed to several investments other than Heroku that had done well, including YC-backed AppJet, Aaron Iba's startup that was sold to Google, and Auctomatic, which was Harj and Kulveer Taggar's startup. Auctomatic fell in the category that Buchheit called "smaller but still nice (2x–3x) exits." He wrote, "As much as I'd like to get a nice 10,000x exit, I'm certainly not going to complain when someone doubles my money!"

Buchheit emphasized that angel investing was not a reliable way to make money—in fact, investors should assume they would lose money. It was far better, he said, to invest in startups with the twin goals of learning about investing and helping the companies. He said, "Anyone doing it with the idea that they can simply find the next Google, invest a bunch of money, and then get super rich is going to be very disappointed."

Buchheit has done well with his investments, but the chances that the value of any given investment will drop to zero are quite high. Congress bars members of the general public from investing in startups like YC's, which don't make their basic financial information public. On Demo Day, the audience will consist entirely of individuals with a high net worth or annual income, what the SEC calls "accredited investors." This originates in the Securities Act of 1933, which seeks to protect unsophisticated investors from losing their life savings. The law requires formal disclosure of essential financial information before stocks, bonds, or other securities can be issued and offered for sale. It does waive disclosure requirements in a few cases, however, such as when the offering would be limited to accredited investors who, presumably, are sufficiently sophisticated that they can evaluate the offering without the enforced disclosures required by the SEC.[11]

Another year would go by before Congress opened a new path to public financing of startups when it passed the JOBS Act (Jumpstart Our Business Startups). This included provisions permitting startups to raise seed capital through crowdfunding, selling small amounts of stock to any and all individuals who were interested in investing.[12] It was not a complete overhaul and it did not affect the way YC companies went about presenting to qualified investors. But it did show that startups, as a category, had become darlings in Washington. The one initiative that could draw broad bipartisan support—the JOBS Act was passed by wide margins in both the House and the Senate—was legislation that helps startups.

19

FIND A DROPBOX

"*Quiet! We're starting!*"

Even with the help of the hall's sound system, Paul Graham's shouting can barely be heard above the noise of the two hundred visitors who fill YC's hall. The first Demo Day has arrived. The guests are squeezed tightly together on folding chairs that have taken the place of the tables and benches. Outside on this late August day, the temperature is already in the eighties and will climb another ten degrees. On the roof, three industrial air-conditioning units—two of which were added in preparation for this day—are laboring. But they cannot handle the combination of summer temperatures and the heat produced by the mass of bodies packed into the main hall. Inside, it's not comfortable.

The attendees are partners at venture capital firms or tech company founders or executives who are retired and now doing angel investing full-time. YC is now known well enough among high-net-worth individuals with a special interest in tech startups that today's Demo Day has drawn a husband-wife pair of celebrities from the entertainment world. Demi Moore comes in first, and many minutes later she is joined by Ashton Kutcher.[1] YC founders, who are banished to the outside until their individual turns come, set upon Kutcher when he arrived—they knew he was coming—and tried to introduce themselves and their startups, scrabbling to get his attention. With difficulty, he had finally gained sanctuary at the front door.

"All right," says Graham after the crowd quiets. "Welcome to the thirteenth Demo Day. If you've been here before, you probably noticed the room is bigger—again. Y Combinator itself is bigger, too. This time, we have sixty-three startups. They're grouped in three chunks of twenty-one apiece, with bathroom breaks in between. The bathrooms are through that wall. There's one Porta-Potty outside for the adventurous."

Graham invites the attendees to use a designated page at YC's Web site where they can indicate which startups they would like to speak with afterward.

"So, I just want to tell everybody explicitly that we didn't become less selective this batch," he says, explaining that he has heard speculation along this line. "We funded 3 percent of the applicants, just like we always do." He says that the number increased because YC is now taking a larger percentage of the total pool of startups. But he does not mention another possibility: that the startup population itself has been growing as more individuals than ever before decide to try starting a startup.

"It is sort of a schlep that you guys have to sit through sixty-three presentations," says Graham. But he argues, "It's much more efficient than it would be if they all came to have initial meetings of one hour apiece with you at your offices." He tries a bit of warm-up humor. "So, as this function continues, you won't have to go to work at all. You can just come here twice a year." This gets a few laughs. Graham looks down at his notes. "Jessica wasn't sure if I should make that joke."

This is the day on which investors start a YC treasure hunt. Graham sounds excited. "All right. Out of sixty-three, statistically, there's probably an Airbnb or a Dropbox in there. The question is, just which one?"

The allure is not merely that one of these startups may grow quickly into a star with a multibillion-dollar valuation, but that it is still early enough that investors have a chance to put money in under terms that are far more favorable than will be the case when the rest of the investor world recognizes the star startup's value.

The entire summer batch gets the opportunity to pitch all of these investors, or at least to pitch most of them. Some attendees leave before the end, a dismaying sight to those startups assigned to the last slots in the day.

But when the presentations run again tomorrow for a second group of investors, the order will be reversed and the startups that go last today will go first tomorrow. Graham has placed the two most riveting presenters in the batch at the very beginning and at the very end. Chris Steiner, of Aisle50, will open today, and Tom Lehman, of Rap Genius, opens tomorrow.

Steiner introduces his team: "That's George, that's Riley, I'm Chris. We're Aisle50. Which is, in short, Groupon for groceries." His voice is as smooth as yogurt and he manages to sound utterly reasonable even when he ignores Graham's strictures and works in mention of "trillion."

"We will be the premier way for food manufacturers to market their products. The consumer packaged goods industry in the United States is"—he slows to emphasize—"a two-*trillion*-dollar business." He projects logos of Procter & Gamble, Kraft, General Mills, and Kellogg's, which spend $35 billion a year on marketing. "More than *half* of that money ends up in these things." He holds in his upraised hand a sheaf of pages that are easily recognizable as a newspaper's coupon inserts. He pauses a beat. "Freestanding inserts. FSIs. More than 250 million of these go out to newspapers every single week. The problem, of course, is that most people don't look at these. Even worse, newspaper circulation has been declining since 1987. So there's this giant pot of money that's trying to migrate from these old paradigms over to digital."

What companies like Coupons.com have tried to do, Steiner says, is take the entire insert and "jam" it into a couple of Web pages. Kraft and other consumer product goods companies "don't really want a thumbnail image of their product next to all their competitors. What they want is a pedestal, with their product by itself. That's why they're still willing to spend $500,000 to $800,000 for a single-page ad in an FSI that goes out across only half of the country."

Now these companies will have an alternative: Aisle50, which promises to give these companies exactly what they want. "We take their product and put it on a big stage. All alone. With custom copy and great photography. We e-mail the deal out to our users every morning. They buy from us and we load it to their grocery store loyalty card."

This will mean an end to consumers being forced to flip through the newspaper in search of a good deal. That will come directly to their inbox, only one Aisle50 offer a day.

Aisle50 has launched only the week before, with a grocery chain in North Carolina. Other chains have been added swiftly, including a deal signed the previous Friday with SuperValu, the nation's third-largest grocery conglomerate, with 2,500 retail stores like Albertsons and another 2,200 independent grocers that it supplies. Since launching, Steiner says, Aisle50 has been fielding unsolicited inquiries from consumer product companies rather than chasing partners itself.

It is a strong start to the day's presentations, but that is to be expected, given Aisle50's polished presentation the week before at the rehearsals. More surprising is the way that the other startups have greatly improved since a few days earlier. The rehearsing and revising have smoothed the roughest presentations in the batch. Everyone's has been buffed to a high gloss and the group now has a consistency in appeal that was not present before. This will force investors to pay more attention than would be necessary were the quality visibly uneven.

The consistency also renders moot the distinction that at the beginning of the summer had seemed to loom so large, the one separating the startups that had launched by the start of YC and those that had not. Aisle50 was not able to open its site before mid-August, and yet here it is, only one week later, appearing to be on the verge of attaining retail partnerships that extend from coast to coast.

The audience remains unaware of which startups discarded their initial ideas and started anew with an unrelated idea. TapEngage's previous life as Splitterbug could never be guessed as Matt Holden describes TapEngage's ad network for tablets and runs through a demonstration, showing a visit to the *New York Times* Web site on his iPad. He does not have a hockey stick graph to show for TapEngage's nascent service, but he has one that compares the iPad's growth immediately after its launch to that of the iPod and the iPhone—the iPad's has the steepest climb of the three.

In the spring, Brandon Ballinger had come in for his finalist interview as an involuntarily single founder with an uninspiring idea to do an app for neighborhood-level events, like happy hours at nearby bars. Yet today, the replacement cofounder, Jason Tan, is presenting an entirely different, and pretty terrific-sounding, idea for Sift Science, which addresses what Tan describes as a $4 billion a year problem of online fraud. "We're building fraud detection as a service," says Tan. "I want to tell you guys about peer-to-peer fraud. If you see Airbnb, you see a great business. When organized crime sees Airbnb, they see an amazing vehicle for laundering money." He walks through how a criminal creates fictitious accounts as both a seller and a buyer to carry out the fraud. Two YC companies—Airbnb and Listia—are using Sift Science's service.

"This is a real example of a fraudster on Listia," says Tan, showing a diagram. He speaks with the swagger of a seen-everything detective. "This guy in the middle—let's call him Joe—he created all these fake items and all these fake users. He then bid on his own items, just so he could funnel virtual credits into himself. What a *jackass*!" The audience laughs, but Tan is not done. "Our system found Joe. And we're going to find anyone else who tries to *fuck* with our customers!" The choice of verb draws extended laughter in the hall; someone even starts to clap.

Tan resumes. "Now this is a really hard problem, but we're the right team to solve this. My cofounder, Brandon, was most recently the tech lead for Android speech recognition—that's a really hard machine-learning problem. I myself have worked at three startups and was CTO of BuzzLabs, acquired by IAC in April."

✦

Outside, the founders await their turns to present. They mill about, exchanging tidbits of news that dribble out from the show inside. These hours spent standing outside anxiously will turn out to be the time when members of the batch build bonds in a way they had not during the summer itself. Kulveer Taggar would write later of Demo Day, "All of a sudden you get to know almost everyone pretty well."[2]

At one point, a man on a Harley-Davidson motorcycle rides up and complains in a loud voice about a car that is parked on his nearby property. His is a nondescript building used for some kind of light industrial purpose, like all the others in the neighborhood. Late in the morning, its parking lot was empty. A Demo Day attendee must have thought that the property was unoccupied and parked there. That was a mistake. Now the biker is saying that he will put the car into the middle of the street himself. He roars off and then returns to deliver more threats.

The founders regard this drama with detached amusement. They have followed Paul Graham's instructions to park at a public lot situated a long walk distant, leaving the street-side parking spaces for the investors. At this moment, the founders are glad that they parked where they did, safely out of harm's way.

At the first break, the attendees pour out of the doors to stretch their legs, make phone calls, and chat with founders. Many of the founders head inside in vain pursuit of a refreshing blast of air-conditioning.

✦

A week earlier, at the first rehearsal, Jason Shen of Ridejoy had spoken tentatively and failed to emphasize his main points. Today, he is a different speaker, one who exudes confidence. As he approaches the dais, the former collegiate gymnast does a cartwheel.[3] "I thought you guys need a little pick-me-up or something," he says, and goes into his pitch. "Ridejoy is the community marketplace for rides. If you're going on a trip, you can list extra seat space in your car. And if you need to get somewhere, you can find a ride using our site." He explains that Ridejoy is adding an element of "reputation" to ridesharing, a mechanism for payments, and a "great user experience." Shen anticipates a question that may be in the minds of the audience members: "Maybe this is some kind of crazy San Francisco hipster thing. It's not." He shows a photograph of an older man. "That's Michael. He's a former director of Merrill Lynch. And he hosted two plane rides on our site, which just got filled up. And it's not just old guys with planes, either"—this gets a laugh, perhaps because there are old guys with planes who are sitting in the audience. "Our users are about thirty-six years old, on average, and about half are women. Just like on Craigslist, there are about as many people offering rides as requesting them, so it's a balanced marketplace."

Another speaker, Tikhon Bernstam of Parse, had failed to shine at the first rehearsal. Today, however, he gives a model presentation. "We are Parse, and we are Heroku for mobile applications," he begins. "We let you add a back end to your mobile app in *minutes*, without writing *any* server code." He gestures at a slide that illustrates the way apps are developed today. "It *sucks*! Every app developer is rewriting all of this stuff, over and over again, for every single app. *This. Is. Insanity.*" The solution, he says, is Parse, sparing developers from having to give servers or server code another thought. Parse is cross-platform, supporting both Android and iOS apps for the iPhone and iPad.

Bernstam mentions that the company started only eleven weeks earlier yet has managed to complete the code necessary to launch, has opened the service to two thousand developers in private beta, and counts Hipmunk and Weebly—both YC companies—as its clients.

"Our investors see this traction and this trend and that's why we've been able to raise over $1.4 million already from top-tier venture firms, seed groups, and a roster of awesome angels, including some of you guys in the audience," says Bernstam. The figure of $1.4 million, raised before Demo Day, is astounding; no other startup in the batch has raised anything close to that.

Vivek Ravisankar and his Interview Street cofounder have had to work all summer on separate continents—his cofounder in India never succeeded in securing a visa. Today, however, Ravisankar betrays no signs of experiencing any difficulties. "What do you think is the most *valuable* asset in Silicon Valley?" he asks the audience, pausing a long while before answering. "Great programmers! There is a war for talent. And you would have noticed this, even in your portfolio companies. Everyone wants to find and hire the best programmers. How do you solve it? InterviewStreet.com. We have built this platform to help companies find and hire the best programmers."

He shows how Interview Street's Web site offers programming challenges, drawing "thousands of hackers" who attempt to solve the problems. Once they do, they can indicate which companies they would like to work for. Interview Street passes on the scores to those companies, to the companies' delight. "The companies love us because they get the best hackers to interview. The hackers, because they are addicted to programming challenges.

"It's been just eight weeks since we launched," he says, "and we're already working with some of the big companies, like Facebook, Amazon, Zynga. This market is hot and huge." Interview Street is doubling its revenues every month and will soon reach $100,000 in bookings. It's already profitable, he says. Here Ravisankar does some arithmetic out loud that quickly gets to a very large—hypothetical—revenue number. "We get $10,000 for every programmer that is hired. If we match 10,000 programmers, we are going to make $100 million in revenue, and this is just the start."

Another member of the batch, Verbling, a Web site for practicing a foreign language with a native speaker through live video, also invites the audience to imagine an enormous business. The site launched a month ago. Two times each day, Spanish speakers studying English and English speakers studying Spanish are invited to visit the site for video chats. They are paired up and then speak in one language for a set amount of time, shown by a countdown timer, and then switch languages and speak for the same amount of time using the other one. Verbling's founders are Mikael Bernstein and Jake Jolis, two Stanford students, and Fred Wulff, a hacker who left his position at Google to join them at the beginning of the summer.

"How many of you have ever tried to learn a foreign language? And how many of you were successful?" opens Bernstein. "If you're like the average language learner, probably not many. It's probably because learning the language alone is tedious, and really, really difficult to find the native speakers to practice with. We solve this problem."

He says that eighteen thousand people have signed up on Verbling and another nine thousand have signed up for languages that have not yet been offered. When he talks about the potential business, the figures become quite large. "We are a huge market: $32 billion is spent on self-paced language learning alone," he says. "There are nearly one million American college students just learning Spanish. There are one billion people learning English, all around the world today." Verbling is offered free at the moment and in the future will let supply and demand for different languages set varying prices for the service.

Most of the teams provide a sentence or two about their professional qualifications at the end of their talk. Bernstein's past is unlike anyone else's in the room, however. "I served as a Swedish-Russian military interpreter and interrogator in the Swedish special forces." The audience laughs at the novelty of hearing "interrogator" and "special forces" here. He continues: "Jake has translated for the United Nations. And Fred is a Stanford CS alum who left Google for Verbling. Between the three of us, we speak six languages fluently."

✦

The second break begins and some investors have decided they can absorb no more presentations and are departing. Ashton Kutcher and Demi Moore are among them, but again outside Kutcher needs a security detail, which he does not have, to slice through the crowd of founders that collapses upon him. He shakes free, pleading, "Have to catch a plane for Letterman."

When the break is over and the seats reoccupied, Paul Graham stands in front and says, "If you did not hear the announcement about the car that is being towed out into the street back there, if you have a black Mercedes AMG, your car has just been dragged by some *nut* with a forklift and a rope out into the middle of the street. Sideways. So if that is your black AMG, go get it." The car turns out to belong to a billionaire whose cars, it can be assumed, are not usually placed in close proximity to a biker eager to use a forklift as an instrument of class warfare.

Graham redirects attention to the remaining program. "OK. Ready? Last batch! Afterward, we take away all the seats, this turns into a reception. You can hang out and talk to the founders for as long as you want." He tries another joke. "We'll have copious amounts of drink—a great opportunity to decide valuations."

Clerky, whose two cofounders, Chris Field and Darby Wong, are both attorneys and hackers, gets the full attention of the room with Field's opening line: "Clerky is making lawyers obsolete." Someone in the crowd shouts out, "Yeah!" Field describes how Clerky's software drafts documents, collects signatures, and keeps the paperwork organized. "Software is eating the world," he says, invoking the phrase used in Marc Andreessen's *Wall Street Journal* essay that appeared three days earlier (Andreessen himself is sitting in the audience).[4] "The automation of routine legal work is inevitable. So why hasn't it happened yet? You need founders who are lawyers and founders who are engineers. If you only have lawyers, you end up with crappy software, serving the lowest tier of the market"—the slide shows LegalZoom at the bottom—"leaving the rest wide open." Field says $250 billion is spent every year on legal services, but does not claim to be addressing all of it, only the 8 percent he says that is so routine that it can be automated. By starting with a humongous number and then showing restraint by targeting only a relatively small slice, he ends with a figure—$20 billion—that is still enormous but seems perfectly credible.

Clerky "is the biggest thing to hit the legal industry since e-mail," says Field. "But the cool thing is you can use Clerky today to invest in any of the sixty-three companies here today. You can close your investment in minutes on your phone or iPad. And if your phone isn't smart enough, come find me and borrow mine. We're Clerky, and we're making lawyers obsolete."

Near the end of the day, Codecademy's turn arrives. The investors do not know that the week before, at rehearsals, Zach Sims had little to say that seemed compelling. The Web site had not launched, so he had no graph of user growth to show. Nor could he articulate why Codecademy would grow into something larger than what it seemed to be, a modest project with limited appeal. Today, however, Sims's presentation bears no resemblance to the earlier version.

Four days earlier, Codecademy's site had launched without intending to do so. Sims says that the site has attracted 250,000 users—not merely visitors, but visitors who have taken the trouble of completing a Codecademy exercise, getting started on learning how to code. Most users are spending more than an hour at the site at a time—and this is when it has nothing much to offer quite yet, just eight simple lessons on JavaScript basics.

Codecademy had only to open its doors, and the world has come to it, fighting to get through the doorway. Groupon and Microsoft want to offer lessons about their programming interfaces. More than two hundred coders have expressed an interest in writing programming lessons. Bilingual volunteers around the world have offered to translate lessons into their native languages. Within a month, Sims promises, Codecademy will have two hundred times the content it currently has.

Sims says that a number of companies have contacted Codecademy, asking if they can get in touch with users who finish the exercises. "We're going to be more than happy to take their money in the future."

Of the sixty-three presentations, this is the one that is most perfectly aligned with the Zeitgeist. *Software is eating the world* and Codecademy is survival school. At the moment, its Web site offers nothing much beyond a placeholder for the school that has yet to be built. But it offers hope that programming skills can be acquired, conveniently and painlessly, by those who do not code, which is to say just about everyone. Investors quiver with anticipation at backing a company that will sell to nontechnical people, the normals, who fear the future will leave them behind.

CampusCred gives the next-to-last presentation—"this could be a $5 billion a year market"—and then Rap Genius ends with a flourish—"there hasn't been another startup in Y Combinator *history* that's had *this* much traffic on Demo Day."

"We're done," Graham announces. The chairs are cleared and the schmoozing begins. The founders circulate, looking at name badges, searching for the investors who used YC's online system to indicate an interest in meeting with them. The investors talk, detach, and move a few steps, searching for particular founders. It's not unlike trying to find a romantic interest at a middle school dance, inching forward in a packed gym.

It's a vestigial ritual and nothing more. If investors and founders fail to find the ones they seek, they will get in touch with one another by e-mail and phone calls in the next few days and the fund-raising process will extend weeks, in many cases, months, for most of the companies. But to the founders especially, it seems important at the moment to be able to meet prospective investors in person before they disperse. Physical presence still seems more substantive than a virtual one.

"It's hard to say exactly what it is about face-to-face contact that makes deals happen," Graham observed a few years ago, "but whatever it is, it hasn't yet been duplicated by technology."[5]

20

DON'T QUIT

The sixty-three startups now must find their way along sixty-three separate paths. They were not plunged into loneliness. The YC model does not have a shared work space, so even during the summer the founders spent most of their time working in their respective apartments, separate from the others. They are as self-sufficient a group as you're likely to find in the Valley. But with the end of YC, a mental adjustment to the new circumstances is required nonetheless. Ahead is the slog of indeterminate duration, where the lighting is less bright and there will be no jolts of invigoration provided by seeing peers at the weekly dinners.

During the summer, everyone was fully abreast of news about everyone else. After YC, news does not circulate as efficiently, and some founders take pains to keep disappointment to themselves. The facade of a company's Web site or the presence of its mobile app in the app stores conveys solidity that may not necessarily exist. Some have already decided to abandon the idea that has just been presented to investors at Demo Day and are searching for another one. Those that experience immediate setbacks and struggle to find their footing do so privately, out of public view.

About 30 percent of YC companies, sooner or later, will experience a split among cofounders. Fissures in relationships among some of the teams in the summer batch have already opened up. Ben Pellow, CampusCred's resident hacker, has left the startup. Paul Chou has left Opez. NowSpots' cofounders, Brad Flora and Kurt Mackey, who had met only just prior to applying to Y Combinator, have also parted ways. After the split, Mackey, who had been the hacker member of the NowSpots pair, joined the MongoHQs as a new cofounder.

At the kickoff meeting, Graham had told everyone that he expected more than half of the startups in the batch would fail, a prediction that at the time seemed extremely remote. It does not seem so remote now that the startups have been pushed out of the nest. None will disappear as quickly as one in the first batch, in 2005, whose founders at the end of the summer simply turned off the lights and returned to graduate school. Today is a different era. Many members of the batch have raised hundreds of thousands of dollars—and even the ones that did not raise additional money have the $150,000 from the Start Fund and SV Angel—so all will last a while. But when death comes, it will likely arrive in the same gradual way it did for early YC companies, starting with a long stretch in which the product goes without updates.[1] The founders leave to do something else. The Web site might remain live but no one is at home.

Chad Etzel, a single founder when he was funded by YC in the winter 2010 batch, wrote a post the following year about his post-YC experiences, titled "Startups Are Hard." He said he wanted the post to serve as a corrective to the sunny stories that make it appear that "anyone can suddenly decide to be a founder and the next week find themselves swimming around in a kiddie pool full of angel/VC money." He and Paul Stamatiou, the cofounder who joined him after YC, largely failed to raise money for their product, Notifo, a notification service for technical people, such as systems administrators who need to know when a server goes down.[2] The experience of being turned down by investors when batchmates seemed able to raise large sums with ease was so crushing that "there was literally a week when I would just stay in bed all day so I didn't have to face the world. This was the lowest point in my entire life."[3]

In June 2011, Stamatiou went on to cofound a new startup, Picplum, a member of YC's summer batch. In September, Etzel announced he had to stop work on Notifo and seek full-time employment to meet his living expenses. He said he would not immediately shut down the service but he warned users not to depend on it, urging them to move to alternative services.[4] Notifo's Web site looked alive but no one was there.

✦

A week following Demo Day, the founders come to YC for the final dinner. They are in no mood for looking back and reminiscing. Most are in the midst of fund-raising, a subject that holds them in an obsessive grip. The evening program for this final dinner consists of Paul Graham and Sam Altman offering fund-raising advice, with the hope that the founders will soon get back to work on their products.

The founders have been busily exchanging rumors about terms of investment offers. "*Do not* chase high valuations!" Graham instructs at the front of the hall after dinner. "How many times do I have to tell you guys this? Fund-raising is not the battle that you're fighting here. The battle you're fighting is to make a successful *company* that makes this great *product* and gets tons of users. And fund-raising is just this tedious errand to be got over with as quickly as possible, and I mean you, Rap Genius." All of the founders laugh.

Graham tells a cautionary tale about a YC startup in an earlier batch that became intoxicated with fund-raising. One of the founders, a gifted deal maker, persuaded a well-known venture capitalist to increase significantly the valuation that was initially offered. This led to a chain of events in which the founders left the company because they ended up owning so little of it that it was no longer worth their hanging around.

When Graham finishes the story, someone who is from Europe calls out, "What happens to those founders? Are they going to raise money again here in Silicon Valley?"

"They're off doing new startups."

"So if you raise money and you screw up, then you—" The audience's laughter drowns out the end of the sentence.

"Silicon Valley is very forgiving of failure," says Altman.

"Tell us if something big happens," Graham tells the founders. "Don't conceal it. You don't have to worry—we're not going to ask for the money back." This is now an old joke but it still produces a laugh. "If one of your two founders is leaving—or other disaster—tell us! We've probably seen it before."

Even as one batch of founders is about to depart, it won't be long before YC is receiving applications for the next batch. Near the end of the evening, Graham asks the summer's founders not to be overly specific in the advice they dispense to applicants.

"If you know some trick, something that someone could put in their application that would make us think, 'Oh my god—we've got to interview these people!,' please don't tell them," Graham pleads. "Every time at interviews we get more and more people who say things that are just *exactly* what we wanted to hear. We know the alumni are telling them, so don't do that."

"Have you ever considered changing the length of YC?" asks someone in the audience. "How did you stumble upon three months?"

"We stumbled upon three months because it was originally a summer program and it turned out to work, so we didn't change it. Have we ever thought of changing it? Not really. I don't think you could make it that much shorter, and making it longer wouldn't help much. So we accidentally got it right the first time."

Another founder asks Graham how well the expansion of the batch, from forty-four startups in the preceding one to sixty-three in this one, turned out.

"The first dinner, I walked in here and I thought, 'Jesus, there's a lot of them.'" The group laughs. "Then I realized I had thought that five times before; everything had turned out fine." He says he didn't really know until Demo Day whether the summer's experiment had been a success. That's when YC alumni who saw the pitches told him that they thought this batch was as good as, or better than, earlier batches. "So expanding turned out all right. A lot of things broke. But a lot of things always break. So we're going to fix the things that break." He tries out a joke: "And then more things will break. Until I get old and die."

Graham warns the founders that they can expect to feel an emotional letdown. "Everybody rides really high on Demo Day. And there's like a huge plunge off a cliff, into a trough of depression after YC." The group laughs, not realizing he is serious. "There is," he says.

"Reunion dinner on October 4!" Jessica Livingston calls out from the side, referring to the dinner that comes in five weeks for those founders who are still in town. No college offers to organize a reunion mere weeks after commencement, but the YC dinner provides a way to postpone the dispiriting isolation that will descend upon many founders.

✦

The day of the reunion dinner arrives. For some startups, fund-raising continues, but for most, it's complete and the founders have turned their attention back to their products. Many are also hiring their first employees. They had heard from YC alumni that the startups that have been the best at hiring found all of their new hires outside of Silicon Valley and moved them here. Best friends will relocate to be with one another, but employees are not best friends and their weighing whether to accept an employment offer and move to a place with a notoriously high cost of living will rely on an economic calculus.

Before Demo Day, Felix Shpilman, Yuri Milner's associate, had suggested one way a startup might lure prospective employees to the Bay Area: by offering housing—not a housing allowance, but actual housing. He told one company, in a manner that suggested he was serious, that it should find a large apartment building and lease many units. One would be for the office. Several others would be for the founders. And other units would be for every employee who moves to the Bay Area. Everyone would have their own personal space and no one would waste time commuting. Startup heaven.

✦

"A lot of you, the toughest bit is coming up right now," Paul Graham says, opening the reunion dinner's program. "When you're, like, launched off the end of the carrier deck and don't yet have sufficient airspeed"—some in the audience laugh—"and the water is only seventy feet down." He tells everyone not to feel demoralized. "This early stage in the startup's life, everyone runs out of morale," he says.

He anticipates that the founders are feeling panic at this very moment, convinced that users do not like their products. "Don't think"—he changes his voice to a high register—" 'Oh, users don't like our product! We give up!' " Instead, founders should proceed analytically and unemotionally: " 'Users don't like our product? OK, we'll analyze it. We still have lots of money in the bank. We'll figure out something.' "

Graham urges the founders to stay in close touch with YC and with one another to ward off postgraduation depression. "The only thing that has to stop is Tuesday dinners. Everything else can keep going. You guys can hang out with one another just as much as you ever did. You guys can come to office hours just as much as you ever did."

The founders understand, however, that the way YC works, centered as it is on the physical presence of participants, the formal end of YC does mark a significant change. If they are not located in the immediate vicinity of YC in Mountain View, they are not going to visit regularly for office hours.

Most of the founders in the summer batch who came from another place have returned home. Rap Genius is back in New York; Snapjoy, in Boulder; Nowspots and Aisle50 in Chicago; TightDB in Denmark. The MongoHQs did decide to move back to Birmingham. But their newly added cofounder, Kurt Mackey, has rejoined his family in Chicago, where he works remotely. At MongoHQ's office in Birmingham, enormous monitors mounted on one wall and Skype are used to give Mackey in Chicago, and another hacker in Portland, virtual presences in Birmingham during the workday.

✦

One day, near the end of the summer, the Graffiti Labs founders ruminated about the social dynamics among the founders in the very large summer batch. Mark Kantor and Tim Suzman remembered the earlier time when they lived in the "Y Scraper": a twelve-story building on Taylor

Street in San Francisco that acquired the nickname because of the large number of YC-funded founders who lived and worked there. It was one of the only places in the city that offered furnished apartments on a month-to-month basis. At one point, thirteen YC-funded startups were there, including Justin.tv, Weebly, Dropbox, Xobni, and Scribd.

The experience for YC companies then was different than it is now, Suzman said. "In the early batches, everyone in YC knew each other and were tight friends. They were all up to date on what each company was doing." They spent their free time together, too.

With the much larger size of the summer batch, the founders got to know not just a smaller percentage of fellow founders, but fewer founders, period.

"I actually think they should do things like assigned seating at the beginning of dinners," said Kantor.

"I don't know," said Ted Suzman. "It would feel infantile."

Kantor said he didn't know many of the 160 founders of the summer batch. "There aren't that many I would call on the phone right now. I think there's something missing there."

"And a lot of them seem like people we would want to know," said Tim.

"The question for PG and company is, does this matter?" asked Ted.

Perhaps the trade-offs—a much enlarged professional network that comes at the price of diminished closeness within the internal personal network—will yield a net gain for the founders. It's hard to separate out concrete loss from nostalgia for the way things were.

✦

Six years after YC had made its first investments, Graham attempted to tote up the value of the fund's investments to date. Excluding the most recently funded startups and looking only at the 208 startups funded from 2005 through 2010, he reported that five YC-funded companies had been acquired for over $10 million each, and twenty more had been sold for less. An acquisition is an exit, and the value set by the buyer at the time of the transaction is indisputable. An initial public offering, a milestone no YC-funded company has yet attained, is the other form of an exit that sets a valuation by objective means.

The values of most of YC's investments are not so easy to determine, however. One can use the valuations negotiated at the time of the most recent financings, but there is no way to know, in the absence of public trading, whether a high valuation used in a private financing will hold up as a predictor of the exit valuation. With these caveats noted, Graham added up the valuations of the twenty-one most valuable YC companies that had not exited and came up with a total of $4.7 billion.[5] Most of that came from a single company: Dropbox. Paul Buchheit points out that in YC's portfolio "the number one company is worth more than the next 199 companies combined, while number two is worth more than the next 198 combined, and so on."[6]

One could say that the one outlier—Dropbox—is the only one that matters to the YC fund. Perhaps it would be better to say that YC is in the hits business and uncertainty about which startup will become the one monstrous hit benefits many founders who are funded, just as, in book publishing, another hits business, uncertainty about future reception in the market benefits many authors who are given a chance.

Graham and the other YC partners tell the founders that startups fail only when founders give up. It is not necessarily in the interest of founders to follow that advice indefinitely, however. At a certain point, surely the personal price paid by founders to attempt to resuscitate a failed startup is simply too high. But then along comes a story like that of Iminlikewithyou (I'm in Like With You), originally a member of YC's second batch in winter 2006. This is a tale of endurance with a happy ending.

When it first launched, Iminlikewithyou was an online dating site. The founder, Charles Forman, decided to convert it into a game site and renamed the company OMGPOP. It raised $17 million from investors, created more than thirty-five games, and yet it could not generate much revenue. It was on the brink of death, mere months from running out of money in the spring of 2012, when one of its smartphone games called Draw Something, which resembled Pictionary, became a breakout hit. It was downloaded thirty-five million times in a matter of weeks and its users quickly created more than a billion drawings. It also began generating hundreds of thousands of dollars of revenue—a day. In March, the game giant Zynga purchased OMGPOP for a reported $210 million, making it the second most successful exit to date for a YC-backed company.[7]

For the members of YC's summer 2011 batch, and for founders of all startups, a tale like that of OMGPOP, with the dizzying swing from near death to champagne celebration in just a few weeks, should provide spirit-lifting sustenance for lean times.

21

SOFTWARE IS EATING THE WORLD

The idea that programming was for everyone pervaded the air, or so it seemed as Codecademy's two young founders, Zach Sims and Ryan Bubinski, hopped from one triumph to the next after Demo Day. They were among the more successful fund-raisers in the batch, securing $2.5 million. Then, on January 1, 2012, they unveiled "Code Year," an inspired marketing idea: Codecademy invited nonprogrammers to make a New Year's resolution to learn how to code. By signing up at Code Year, users would receive free weekly lessons supplied by Codecademy over the course of the year.[1]

Sims and Bubinski placed a story with TechCrunch on New Year's Day about Code Year, but the most important thing they did to spread the word was to make it easy for everyone who signed up to get friends to join them.[2] With a click, a new member could tweet or share on Facebook a message that was already prepared: "My New Year's resolution is to learn to code with Codecademy in 2012! Join me."

Join they did. Within forty-eight hours, one hundred thousand users made the Code Year resolution.[3] Three days later, the total passed two hundred thousand. One of the new registrants was New York City mayor Michael Bloomberg.[4] By mid-January, more than three hundred thousand had signed up and the initiative caught the attention of the White House, permitting the founders to use the phrase "partnership with the White House" in the heading of a post on Codecademy's blog. Such sponsorship had not been anticipated when the founders were scrambling to get a barebones Web site ready just five months earlier.[5]

✦

Justin.tv's founders have decided that its two promising, but technologically dissimilar, experiments—Twitch.tv, video game broadcasting, and Socialcam, the app for sharing videos captured with a smartphone—are best pursued separately. Socialcam has been spun off as a separate company with its own offices. Michael Seibel is its head and two engineers from Justin.tv have gone over to the new startup. Meanwhile, Twitch.tv, overseen by Emmett Shear, continues to grow rapidly. In December, it drew twelve million unique visitors, which is 600 percent more than Justin.tv had attracted to its gaming content a year previously. Shear and his partners have designated Twitch.tv as the future of Justin.tv. It has about forty employees and another twenty will join in the next few weeks, so it will soon be triple the size of Justin.tv a year previously.

✦

It's Tuesday, January 3, the day of the first dinner for YC's winter 2012 batch.

YC has a long-standing tradition of inviting a few alumni to the first Tuesday dinner to give short presentations and field questions from the new arrivals. Seven alumni will speak this evening. Four are from the summer batch: Michael Litt of Vidyard, Thejo Kote of MileSense, Tikhon Bernstam of Parse, and Tom Lehman of Rap Genius.

"Welcome to the first dinner," says Paul Graham, as he prepares to introduce the evening's speakers. "They will tell you what they wished they'd known when they were in your position. But when they literally were in your position, there was *another* bunch of alumni out here who told them the same kind of things they're going to tell you. They didn't listen. So better listen to them this time!" He lowers his voice slightly and looks down, pretending to mutter to himself: "—he said, shouting into the wind."

Before the speakers are given the floor, Graham presents some numbers about the winter 2012 batch. "This is the fourteenth batch of startups we've funded," he says. "Between them all, there are 383 startups. There are eight hundred alumni exactly. There are sixty-six startups in this batch, out of 2,027 applications. Last batch, sixty-four out of 2,089 applications. So almost identical."

Graham does not announce the number of women in the batch or members of other groups that are not "the bunch of white and Asian dudes" who comprised almost all of the summer batch. But there are nine women in this winter batch, up from two in the summer, and four African Americans, up from none. Changes in the numbers from the two most recent batches cannot be read as a trend, but they do show that there have been significant swings from batch to batch.

One founder in the winter batch is Daniel Kan, twenty-five, Justin Kan's younger brother. When he moved to the Bay Area in 2009, he had no interest in starting a startup and said he would never do so. But he ended up working for one and spent a lot of his time with his older brother and Shear and their friends. The environment determines the socially accepted norm. "If all your friends are going to go work at finance companies, then you're going to work at a finance company," says Shear. "That seems like a normal thing to do. If all your friends are doing startups, you're going to do a startup." Daniel had not deliberately immersed himself in startup life, but the effect on him was the same as if he had. Here he is at YC.

And joining him as a cofounder is big brother Justin. Graham announces, "Part-time partner Justin Kan is in this batch with you! He has decided to do another startup." They and a third cofounder have launched Exec in San Francisco, which offers to dispatch, within ten minutes of a request, someone who will do most anything.

"We are delighted beyond words," says Graham. "And we'll now be able to witness, for the first time, the spectacle of someone having office hours with himself." This gets a good laugh.

This will effectively be Justin Kan's third YC batch. Also in the winter batch is Michael Seibel, who has brought Socialcam to YC; this will count as his second time in YC. Graham introduces him and six other YC alumni from previous batches.

Each startup in the batch is again offered $150,000 in convertible notes, but now Andreessen Horowitz is a contributor, alongside the Start Fund and SV Angel.

"Demo Day is March 28," says Graham. "This year we're going to have Demo Day at the Computer History Museum." At the summer Demo Day, the temperature inside the hall had hit eighty-six degrees. "That was not good," he says. Moving Demo Day to another location will also remove the problems with neighbors that had surfaced in the summer.

"There are eighty-eight days left before Demo Day. Now, if there was ever a time to work, it is this. One of the handy things about Y Combinator is it gives you an excuse for dropping out, right? You can tell everybody who wants to bug you, or distract you, you can say, 'Sorry, I'm in this program. They deduct points if I don't start working as soon as I wake up until the time I fall asleep. So I can't do whatever it is.' It's socially excused to work all the time—very convenient."

✦

The evening's speakers present a miscellany of advice. Much of it concerns fund-raising or hiring, two tasks that are relatively remote to the incoming batch. Tikhon Bernstam, however, has some advice about choosing the idea for the startup, which even founders who think the matter is settled may reconsider after hearing what he has to say.

"I'm Tikhon from Parse," Bernstam begins. "If you're doing anything mobile, you should be using Parse." It would be surprising if there was anyone in the hall who was not well familiar with Parse, even those who were not working on smartphone apps. This was the startup that had raised more than any other in the previous batch—$7 million—though Bernstam does not mention this. He completes his self-introduction by mentioning that he had cofounded Scribd in 2006.

"I want to talk about why good, smart, determined founders fail," he says. "Drew, from Dropbox, came in and gave a great talk in our last batch. He made two really interesting points that I don't think get discussed enough. One: you really need a huge market. And two: you need founder-market fit."

Bernstam says that he is aware that these points have been made by many others but he fears that most of the founders present will ignore them. "But it's easy to find really interesting markets," he maintains. Go to the *Inc.* 500 list and see what the fastest-growing companies are. File sharing is an example of a category that is so large that it accommodates many companies. Some are "crappy"—he names a couple—and some are great—he names two YC startups, AeroFS and Dropbox. The very number of startups in the category shows that the market is huge.

"You don't need to be perfect," he says, again mentioning the low-quality file-sharing companies to illustrate the point. He advises, "Just focus on one of three things: One, be cheaper. Two, focus on a niche. Or, three, be 10x better than the other products out there." Dropbox followed the third course, but many companies are trying the other two.

"It's not critical to be first," he says. "Dropbox wasn't the first company in file sharing and backup; Google wasn't the first search engine." (Nor, one could add, was Facebook the first social network.)

"Look at the deal space," he suggests. "Besides all the ones you've heard of, like Groupon and Living Social, there's a ton of them, making a ton of money, and they're mostly crappy clones. Or Codecademy, of the last batch, is a great startup in the education and hiring spaces."

As for "founder-market fit," he suggests that the fit should be self-evident, but sometimes founders delude themselves. Gazing out at the hall filled with hackers in T-shirts and sweatshirts, he says, "Let's take the fashion industry. There may be six of you who are qualified to be working in that space."[6] He recalls that when he was first in YC in 2006, he had two roommates, also in that batch, who were "hard-core hackers working on a fashion startup but probably would have been better off working on something else."

✦

"My name is Mike," says Michael Litt of Vidyard. "We're essentially YouTube for business. Anything you're doing that's video related—everything from video as a marketing tool, creating and playing your video—we could probably hook you up."

When he and his cofounders went through YC, they purposely isolated themselves from all distractions. "We moved to Los Gatos. We got a big house with a pool and a hot tub. We told our entire families, friends, girlfriends at the time that didn't make it through: we were spending the full three months, which ended up being five months, locked down, no communication—erase us from all family events, gatherings, parties, whatever, for that period." He grants that this was severe and not what he would call "healthy." Nonetheless, he says, "It's very important to focus because Demo Day is the best opportunity your company will ever have to present to two, three hundred active investors." He says he has many friends who have done startups outside of YC who have not had the same opportunity to pitch investors.

A founder in the audience asks the panel if they can recall an occasion in which they had ignored what their gut said and yet had a good outcome.

"Actually, it's interesting that you ask that," says Graham, who responds first. "Because when people come and speak at Y Combinator, I often ask them in the Q&A session what they know now that they didn't know when they were in your position, and probably the single most common answer is to trust their gut." His comments are leading away from the point implied in the question, which is that it may be best to ignore the gut when making a decision. Graham says that even if a founder hopefully thinks a partnership with an eminent person or company will assure the startup's future—"they have the magic bullet, they're going to make us!"—the founder may also have unvoiced misgivings. "You should listen to your inner worry," he says. "If you have misgivings about doing something, you should really stop and think about it." The speakers who come to YC who are older consistently say that they wish they had trusted their gut when younger. "I don't think it's really that they have better instincts," he says. "It's now they have the confidence to trust them."

Litt has a story about ignoring his gut, which actually turned out well. "We got e-mailed by Google corp dev just after Demo Day," he says.

Graham interrupts to explain for founders who are not familiar with the term "corporate development" or its shorthand form. "Corp dev are the people who buy people."

Litt continues. "Everybody said, 'Don't meet them.' My gut instinct was not to meet them. But I went and met them. I was very guarded and didn't say anything about what we were doing. I ended up telling one of our investors. And he told someone else, and it spread that we had a meeting with Google corp dev." Before he knew it, prospective investors were mentioning to him, "I heard you turned down an acquisition offer from Google." Litt would say only that "I had a meeting with corp dev, yeah," but that was enough. "It just spread like wildfire and helped incentivize the investors," he says.

✦

After YC did a survey of the startups in the summer batch to see how the group's fund-raising efforts had fared, it became clear that investors had been selective. Mere membership in the YC class did not mean all could raise capital. The Start Fund and SV Angel's $150,000 investment in each startup meant that no company faced imminent closure, but twelve of the sixty-three, or about one out of five, either did not try to raise additional funds or failed to do so. Seven companies that did raise money received small investments, between $15,000 and $60,000.

At the other end, the startups that investors were most interested in did very well. Aisle50 raised $2.6 million; another, $3.26 million; and the class leader, Parse, raised the $7 million. The median amount raised by the fifty-one companies that received investments, separate from the initial $150,000, was $850,000.

How long the money will last will depend primarily on a single factor: whether the startups hire quickly as they try to reach profitability. The longer that the founders can rely only on their own labor, working on the subsistence salaries they pay themselves, the more time they will have to prove out their idea—or, if they must, discard the idea and try another.

✦

The winter batch founders are now about to be entertained.

"What's poppin'! I'm Tom, representing Rap Genius. We're a Web site devoted to crowdsourced explanations of rap lyrics." Tom Lehman is a born performer. Just as he did when he presented at gatherings in the summer, here again he talks and gestures animatedly, holding the attention of the hall.

"Being here for the first dinner makes me a little emotional," he says. "It reminds me of my first dinner at Y Combinator. I was sitting there in the corner, back there—" He points to the far corner, the one most distant from the entrance. "I was thinking, 'I don't know anyone, I'm nervous, I just want to chill with my cofounders.' Since I've been there, I want to kick things off for you guys and break the ice a little bit. So I want everyone to stand up and hug someone who is not your cofounder." The group laughs but does not immediately move. "Seriously, do it!" he instructs. "This is not a joke!"

A miracle transpires: everyone stands up and hugs a stranger, amid laughter.

When everyone sits down, Lehman continues. "That hug, right there, is a metaphor for what you guys should be doing during the opening days of Y Combinator: putting yourself out there and making friends even if it makes you uncomfortable. So if there's the spiky-haired kid and you don't know whether he's your kind of guy, or maybe there's a hotshot startup that's all YC alums and you're kind of nervous to approach them, don't be nervous. Go up to everyone. Everyone's down to be your friend."

Lehman recalls how his batch, like this one, was large; it was impossible to be close friends with all founders. "But you can definitely meet everyone and kick with them a little," he says. "So introduce yourself to everyone and be homies with everyone."

He tells a story of how he became fast friends with one of his batchmates, Geoff Schmidt of Meteor. "He was this really tall guy and he was always wearing this shiny-ass silver jacket. And I was like, 'What is the deal with that jacket? Who is this guy? Is he too cool for school? Is he down to be my friend?' Fast-forward two months later, we're best friends and I'm screaming at him from across the party"—Lehman shouts out in a loud voice, as if it's difficult to be heard—"Geoff! Look!" Lehman tugs at the collar of his Adidas track jacket, which is a woman's and neon green. "I've got a weird jacket, too! We're jacket bros!

"The other reason I wanted you guys to hug is because you should literally be hugging each other more. As cofounders, your relationship is as important or more than the product you're building. So you need to work on that relationship in the same way you work on your product. And the two best ways to work on that relationship are: one, show love, and two, go to the gym together every day. Seriously. Seven days a week. Do it. It's important.

"There was actually a third reason for the hug—it presented an ambiguous situation. I didn't tell you who to hug. You had to find someone to hug, and maybe he's trying to hug someone else, and that makes you a little nervous or whatever. That ambiguity in the hug dimension is a metaphor for the ambiguity you will face with your business in Y Combinator," he says. "You're going to have to be comfortable handling this ambiguity in tough situations because there's no one who can truly help you make these decisions." Yes, Y Combinator can provide the advice of YC partners or of outside experts, and founders may think, "We've got this tough situation, and this person is going to tell us what we need to do and remove the ambiguity." But that is not the case, he says. Office hours do not absolve the founders from making difficult decisions themselves. "If you have a tough problem, where you can see both sides of the issue and you don't know what to do, you're going to have to make the call yourself," he says.

"One problem we had when we were in Y Combinator was figuring out how to expand Rap Genius into other verticals. Rap Genius is for rap music, but we want to explain all kinds of music and all kinds of text, actually. So how do we do that? Well, one thing we could do is have one big site—called Music Genius or whatever. Another thing we could do is have a bunch of small sites: Rock Genius, Rap Genius, Country Genius.

"It's a tough problem," says Lehman. "We went to Paul Graham before one of the dinners and asked him, 'Paul, what do we do?' So he thought for a second and he said, 'OK, here's what you've got to do. You've got to take all of the verticals and put them under one site and call it Definator.com.' Now, 'Definator' is *not* an amazing name for our product." Lehman's reaction to Graham's suggestion draws laughter. "For one thing, you've got all these community members, they signed up expecting to join this site dedicated to rap music, and now it's 'Definator,' Also, 'Definator' sounds a lot like 'defecator.' " This too brings laughs.

"But what did I do the second PG walked away?" Lehman asks. "I bought Definator.com, Definator.net, Definator.org, Definator.co.uk—I bought all the domain names. Why? Because PG has an insane amount of swag. In that moment, when he was telling me what to do, he had so much confidence and conviction in what he was saying, he could have gotten me to buy any domain name." The audience laughs again, anticipating their future conversations with Graham. "The point of this story is that you guys need to have this kind of swag. You need to suck people, and especially investors, into your own reality."

He heads into the conclusion. "So, what have we learned? One. You've got to put yourself out there and meet people, even if it's awkward. Two. You've got to hug your cofounders and love your batchmates. Three. Experts aren't going to help you solve your problems. Four. You've gotta have swag."

✦

Lehman and the Rap Geniuses do not lack swag. Founders that do not come by confidence as easily benefit by hearing Lehman's entertaining homily. And they will have the benefit of three months' exposure to Paul Graham's variety of swag.

Graham has built a school for startups and he and the other members of the faculty will be major influences. But founders who go through YC will also end up being influences in the same way that student peers feature in the college experience. The comforting presence of the batchmates, engaged in parallel struggles to give birth to viable companies, help make what is iffy seem attainable.

✦

Startups, in Graham's view, constitute a revolutionary economic force, equal in significance to the advent of agriculture, the rise of cities, and industrialization. The other revolutions spread worldwide, but the startup revolution does not seem to lend itself to replication. Software does not need local producers the way every region needed its own railroad or electric power grid. Any place can create software companies, but they are more likely to be found at a startup hub, the place that has all of the necessary ingredients.

"I think you only need two kinds of people to create a technology hub: rich people and nerds," wrote Graham in 2006. "They're the limiting reagents in the reaction that produces startups, because they're the only ones present when startups get started. Everyone else will move."[7] He had grown up in Pittsburgh and attended Cornell and could testify that both places had "plenty of hackers who could start startups," but there was no one in the vicinity to invest in them because "rich people don't want to live in Pittsburgh or Ithaca."

Conversely, Miami had a plentiful number of wealthy people, but "few nerds," he said. Nerds congregate in places that host a leading department of computer science—and also places that "tolerate oddness," because "smart people by definition have odd ideas."

Could any city import the resources needed to create a startup hub? Graham took up the question in 2006 and pondered what would make, say, Buffalo, New York, into a Silicon Valley. To Graham, it was strictly a matter of enticing ten thousand people—"the right ten thousand people." Perhaps five hundred would be enough, or even thirty, if Graham were to be permitted to pick them.[8] Three years later, he suggested that a municipality offer to invest a million dollars each in one thousand startups. The capital required for such a scheme should not seem daunting: "For the price of a football stadium, any town that was decent to live in could make itself one of the biggest startup hubs in the world," he said.[9]

Any place that wants to become a startup hub needs to understand, however, that it requires welcoming hackers and their unruliness. Unruliness is also "the essence of Americanness," Graham maintains. "It is no accident that Silicon Valley is in America, and not France, or Germany, or England, or Japan. In those countries, people color inside the lines."[10] In America, too, failure in business is accommodated. Graham has consistently argued that few people are well suited for starting a startup but that the only effective way of determining who does excel is by having lots of people try: "As long as you're at a point in your life when you can bear the risk of failure, the best way to find out if you're suited to running a startup is to try it."[11]

✦

In Graham's view, there is only one best place to start a startup: among countries, it is America, and within America, it is Silicon Valley. It is a view that comes naturally to someone who works and lives at the epicenter of Silicon Valley's startup ecosystem. But this view of the world does not take into account the way that software has outgrown its original boundaries, as a subindustry within the computer industry. It has become a pervasive presence in virtually all industries: *Software is eating the world.* No economic force of such size can be commandeered by coders in a single place.

In many ways, it is Graham's success with his own startup—Y Combinator—that has helped legitimize the notion in places well distant from Silicon Valley that starting a software startup is an appealing proposition to the ambitious young. Accelerators and incubators have mushroomed all over the world in recent years, including in the UK, Ireland, Spain, France, Germany, Finland, Sweden, Denmark, Greece, Jordan, Dubai, China, Taiwan, Singapore, India, Australia, New Zealand, and the Philippines.[12] It was the dramatic drop in the cost of starting a software startup that enabled Graham and his partners to fund a bunch of startups in Y Combinator's first batch in 2005. Since then, everyone else, it seems, has taken notice of this, too.

The technical skills required to write production-quality code are not within the reach of anyone who takes a few lessons at Codecademy. But there is much broader diffusion of software-writing skills today than there was in the 1990s during the dot-com days. Today, college students who are technically inclined are able to develop sophisticated Web or mobile applications with astonishing speed. No one demands that they first spend five to ten years after graduation working in the bowels of a large software company, like Microsoft or Oracle, before they earn sufficient standing to pay a visit to a venture capital office. It is Y Combinator, more than any other institution, that has spread acceptance in tech circles of the idea that age or years of work experience are not necessarily correlated with coding skill.

Some of the founders at Y Combinator–backed startups taught themselves how to code while they were working on their startups. Their examples offer hope to others who would like to try to acquire some programming skills at an age when it will not come as easily as it does to the young. Even if Marc Andreessen's phrase *Software is eating the world* is not at the tip of everyone's tongue, an inchoate sense of software's centrality is widely understood. The warm reception that greeted Codecademy at its debut speaks to a broad yearning to be a participant in, and not remain a passive bystander to, the spread of software everywhere.

All beginnings embody hope, and Y Combinator gives birth to beginnings by the dozens—with the propulsive power of *Software is eating the world* at their backs. This gives the group portrait a most hopeful cast. Like newly minted graduates, the founders step out into the daylight with nothing but possibility ahead.

ACKNOWLEDGMENTS

To begin, I thank Paul Graham and Jessica Livingston. Their willingness to let me observe unimpeded the workings of Y Combinator is what made this book possible.

I expanded their workload with uncountable requests for information, and I also imposed upon the other YC partners and staff members: Sam Altman, Trevor Blackwell, Paul Buchheit, Kate Courteau, Aaron Iba, Justin Kan, Kirsty Nathoo, Geoff Ralston, Renee Robinson, Emmett Shear, Harj Taggar, and Garry Tan. I wish to thank all for their assistance and for welcoming me into YC.

Welcoming, too, were the founders who composed YC's summer 2011 batch. What a forbearing group! They allowed me to hover when their still gestating or just born startups were at their most vulnerable stages. They spared time to talk with me at length when they were pressed with more important matters. They cheerfully allowed me to sit in on their office hours, visit their apartments, and pester them with all sorts of questions. I am indebted to Ryan Abbott, Nik Abraham, Cesar Alaniz, Nick Alexander, Thushan Amarasiriwardena, Chris Auer, Brandon Ballinger, Erik Berls, Tikhon Bernstam, Mikael Bernstein, Ted Blackman, Tom Blomfield, Ryan Bubinski, Oliver Cameron, Brian Campbell, Xuwen Cao, Hamilton Chan, Jonathan Chang, Jason Chen, Paul Chou, Bjarne Christiansen, Peter Clark, Greg Cooper, Umur Cubukcu, Shaun Davis, Matt DeBergalis, George Deglin, Richard Din, Akshay Dodeja, David Dolphin, Michael Dwan, Pradeep Elankumaran, Ozgun Erdogan, Thomas Escourrou, Robert Farazin, Ethan Fast, Chris Field, Simon Fletcher, Brad Flora, Eric Florenzano, Scott Freedman, Calvin French-Owen, Devon Galloway, Rok Gregorič, Sean Grove, Philipp Gutheim, Tom Hauburger, John Hiesey, Amir Hirsch, David Hodge, Matt Holden, Elizabeth Iorns, Omar Ismail, Jerry Jariyasunant, Jake Jolis, Mark Kantor, Harishankaran Karunanidhi, Dan Knox, Nathan Kontny, George Korsnick, Thejo Kote, Lawrence Krimker, Rok Krulec, Anand Kulkarni, Jayant Kulkarni, Kevin Lacker, Ryan Lackey, Andrew Lee, Jon Lee, Tom Lehman, Tony Li, Ilya Lichtenstein, Brendan Lim, Michael Litt, Sean Lynch, Kurt Mackey, Eric Maguire, Taylor Malloy, Nick Martin, Aleem Mawani, Jason McCay, Ed McManus, Ryan Mickle, Byron Milligan, Scott Milliken, Mahbod Moghadam, Shazad Mohamed, Prayag Narula, Darren Nix, Vibhu Norby, Sam Odio, Patrick O'Doherty, Randy Pang, Srini Panguluri, Sumedh Pathak, Ben Pellow, Vivek Ravisankar, Noah Ready-Campbell, Miha Rebernik, Adam Regelmann, Peter Reinhardt, JP Ren, Christopher Roach, Matt Robinson, Dave Rolnitzky, Andy Russell, Borna Safabakhsh, Tom Saffell, Geoff Schmidt, Riley Scott, Omar Seyal, Michael Shafrir, Sagar Shah, Jason Shen, Roee Shenberg, James Shkolnik, Adam Siegel, Zach Sims, Nick Sivo, Thomas Sparks, Paul Stamatiou, Chris Steiner, Scott Stern, Alexander Stigsen, Ilya Sukhar, John Sun, Ted Suzman, Tim Suzman, Kulveer Taggar, Hiroki Takeuchi, Chris Tam, James Tamplin, Jason Tan, Ian Storm Taylor, Jason Traff, David Turner, Sam Vafaee, Long Vo, Ilya Volodarsky, Jason Wang, Kalvin Wang, Jeff Widman, Darby Wong, Geoffrey Woo, Edward Wu, Yin Yin Wu, Fred Wulff, Ben Wyrosdick, Calvin Young, James Yu, Kevin Yu, Chris Zacharias, Muzzammil Zaveri, Ilan Zechory, Kev Zettler, Andy Zhang, Jon Zhang, Paul Zhang, Shlomo Zippel, and Jason Zucchetto.

Outside of YC, I also received help from (ordered by firm name) Enrique Allen of 500 Startups; Jeff Yolen of BuySimple; Brian Krausz of GazeHawk; Brook Eaton, William Gaudreau, and Eric Thomas of mSpot; Duncan Logan of RocketSpace; Mathew Dunn of Say It Visually; Mark Dempster and Greg McAdoo of Sequoia Capital; David Lee of SV Angel; and David Cohen of TechStars.

From the project's beginning, my agent Elizabeth Kaplan has been my guiding light. At Portfolio, Courtney Young was unstinting in the care that she gave the project; her suggestions greatly improved the raw first draft of the manuscript. Then Niki Papadopoulos, who was the other member of the tag team of gifted editors, helped smooth and tighten the manuscript further. Nicholas LoVecchio made the quiet improvements that only a careful copy editor can. Amanda Pritzker and Natalie Horbachevsky worked on the launch with singular enthusiasm. Chip Wass conjured a terrific jacket. And Adrian Zackheim was unflagging in his encouragement.

The manuscript was improved by suggestions supplied by Pamela Basey, Lee Gomes, Gary Rivlin, and Greg Stross. The pages also underwent the scrutiny of Gail Hershatter, who was pitiless, as always, in marking lazy thinking and infelicitous phrasing throughout.

The College of Business at San Jose State University kindly provided financial support for a leave to work on the book. I'm grateful for the way that William Jiang and Barbara Somers pushed bureaucratic mountains aside to clear the way on short notice.

Finally, I thank Ellen Stross, who provided every form of help I needed, including perspective.

APPENDIX

THE SUMMER 2011 BATCH

Adpop Media: Software for product placement in video after filming

Agile Diagnosis: iPad software for doctors, medical students, and nurses to use when diagnosing patients

Aisle50: Daily grocery deals

BentoBox: (*See* Streak)

BridgeUs: Conference call service

Bushido: Hosting open-source apps

BuySimple [withdrew early]: Micropayments system for media publishers
Name changed from Minno

CampusCred: Local deals for college students
After YC, pivoted to work on TheQuad and mobile app software for college classes

Can't Wait: Mobile app for receiving trailers of upcoming movies
After YC, pivoted to work on Clutch.io, offering software tools to speed the development of mobile apps

Citus Data: Fast database software for very large data collections

Cityposh: Casual online games combined with sponsored advertising

ClassMetric: Software for college students to provide instructors with real-time feedback during lectures

After YC, pivoted to work on Segment.io, analyzing user activity on Web sites

Clerky: Software for automating routine legal work

Codecademy: Online courses to learn how to program

Cryptoseal: Security software for servers in the cloud

DebtEye: (*See* SpringCoin)

DoubleRecall: Advertising software that requires users to type a couple of words related to the sponsor's brand before seeing a Web site's content

Embark: Mobile app for mass transit riders

Name changed from Pandav

Envolve: Chat software for Web sites

After YC, pivoted to work on Firebase, which stores data used by real-time online services like chat and games

Everyme: A social network based around the user's address book

Freshplum: Software for determining optimal e-commerce prices

GlassMap: Mobile app for sharing location automatically in the background without draining the phone's battery

GoCardless: Replaces credit cards for online purchases

Graffiti World: A building game, by the creators of Facebook Graffiti

Hiptic: Personal Web sites

After YC, pivoted to work on mobile games

Imgix: A service for storing and preparing images used by Web sites

Interstate: Project management software

Interview Street: Software to help tech companies winnow the best programmers among applicants

Kicksend: Easy file sharing

Launchpad Toys: Publishes software for the iPad for kids to engage in creative play

Leaky: Price comparison site for insurance

MarketBrief: Entirely automated translation of SEC filings into easy-to-read articles

Meteor: Software for developers to manage data that moves between cloud storage and clients

Name changed from Skybreak

MileSense: Smartphone app to enable safe drivers to get lower insurance rates

Minno: (*See* BuySimple)

MixRank: Service that analyzes competitors' online advertising

MobileWorks: Crowdsourcing tasks that are too difficult for computers alone

MongoHQ: Hosts MongoDB databases

Munch on Me: Daily deals at restaurants

After YC, acquired by CollegeBudget

NowSpots: Supplies tweets from advertisers in place of display ads for newspaper Web sites

Opez: Yelp for individual service professionals like personal trainers, hairstylists, and bartenders

PageLever: Software for analyzing the fans that come to a business's Facebook page

Pandav: (*See* Embark)

Paperlinks: Infrastructure for businesses that use QR codes

Parse: Stores data in the cloud for developers of mobile apps

Paystack: Online payment system for kids to use
After YC, pivoted to work on mobile payments

PhoneSys: Call center software for sales teams
Name changed from Pingm

Picplum: Service that sends photo prints to friends and family every month

Pingm: (*See* PhoneSys)

Proxino: Cloud hosting for JavaScript code

Quartzy: Software for managing inventories of laboratories in the life sciences

Rap Genius: A Wikipedia for annotated rap lyrics

Rentobo: Software tool to help landlords find and sign new tenants for rental properties

Ridejoy: Web site for sharing rides

Science Exchange: Marketplace for outsourcing scientific experiments

SellStage: Product videos for Web sites
After YC, pivoted to work on Videopixie, a marketplace for professional video editing

Sift Science: Software to detect online payment fraud

Skybreak: (*See* Meteor)

Snapjoy: Stores and organizes photos

SpringCoin: Software-based debt counseling
Name changed from DebtEye

Streak: Customer relations management software accessed within Gmail
Name changed from BentoBox

Stypi: Collaborative document creation and editing online
In May 2012, acquired by Salesforce.com

Tagstand: Support for NFC (near-field communication), technology in mobile phones

TapEngage: Creates advertising for tablets

TightDB: Software for developers that makes databases superfluous

Verbling: Service that sets up video chats between foreign-language learners and native speakers

Vidyard: YouTube for businesses

Vimessa: App for video voice mail
After YC, pivoted to work on UserFox, an e-mail marketing service

Yardsale: Mobile apps for peer-to-peer commerce within one's own neighborhood

Zigfu: Tools to create games using full-body motions

NOTES

The following abbreviations are used in the notes:

HN Hacker News

HT Harj Taggar

PG Paul Graham (all sources cited are found at PaulGraham.com unless otherwise noted)

TC TechCrunch

YC Y Combinator

INTRODUCTION

1. Independent tour operators have come and gone, trying to eke out an existence driving tourists by the headquarters of the iconic companies in tech. These companies do not offer tours of their premises, however. The one chance to set foot inside hallowed space is at Apple headquarters—and that is only to gain admittance to Apple's company store to buy official Apple T-shirts and caps. See Mike Cassidy, "Silicon Valley Tour Travels Rough Road," *San Jose Mercury News*, October 24, 2011.
2. Marc Andreessen, "Why Software Is Eating the World," *Wall Street Journal*, August 20, 2011, http://online.wsj.com/article/SB10001424053111903480904576512250915629460.html. Andreessen devotes the essay to illustrating the phrase. Andreessen's partner, Ben Horowitz, mentioned the phrase briefly two months earlier in an essay that challenged the contention that "we are in a new tech bubble," in an online debate hosted by *The Economist*: Ben Horowitz, "Against the Motion," *The Economist*, June 14, 2011, www.economist.com/debate/days/view/710.
3. PG described his reaction to the Y Combinator function when he first encountered it: "You wouldn't necessarily have expected such a thing to be possible. We named the company after it partly because we thought it was such a cool concept, and partly

as a secret signal to the kind of people we hoped would apply." Carleen Hawn, "The F|R Interview: Y Combinator's Paul Graham," Gigaom, May 3, 2008, http://gigaom.com/2008/05/03/the-fr-interview-y-combinators-paul-graham/.

4. PG, "Great Hackers," July 2004, www.paulgraham.com/gh.html.
5. PG, "The Word 'Hacker,'" April 2004, http://paulgraham.com/gba.html. Steven Levy's *Hackers: Heroes of the Computer Revolution* (Garden City, NY: Doubleday, 1984) extricates the word "hacker" from derogatory associations. To Levy, "hacker" simply means "those computer programmers and designers who regard computing as the most important thing in the world." He traces the hacker culture back to the Tech Model Railroad Club at MIT in the late 1950s.
6. "Airbnb Celebrates 1,000,000 Nights Booked!" Airbnb blog, February 24, 2011, http://blog.airbnb.com/airbnb-celebrates-1000000-nights-booked. The company launched its service in August 2008, before it was funded by YC.

CHAPTER 1: YOUNGER

1. http://jasonshen.com/. In early years, the blog had the longish subtitle of "A Blog on Conquering Fear, Doing Great Work and Making Things Happen." It evolved and by early 2012 it took a pithier form: "Conquer Fear & Do Epic Sh*t."
2. In March 2012, YC announced that for the summer 2012 batch, it would experimentally accept applications from teams that had no startup idea whatsoever. The announcement said, "A good startup idea is simply a significant, fixable unmet need, and most smart people are at least unconsciously aware of several of those. They just don't know it. And we now have lots of practice helping founders see the startup ideas they already have." "New: Apply to Y Combinator without an Idea," YC Web site, March 13, 2012, http://ycombinator.com/noidea.html.
3. In March 2012, Loopt announced its acquisition by Green Dot for $43.4 million cash.
4. PG said negotiation was not permitted because "the most significant thing we offer, our advice and connections, we offer a constant amount of to everyone." PG's comments in discussion of "'If people turn us down,' PG says, 'as far as we're concerned they've failed an IQ test,'" HN, May 13, 2007, www.hackerne.ws/item?id=550170.
5. Hawn, "F|R Interview."
6. PG, "Hiring Is Obsolete," May 2005, http://paulgraham.com/hiring.html.
7. A third person, Matthew Fong, also was a cofounder when the Kiko application was submitted but decided a few days later not to pursue the project.
8. Justin Kan, "My Y Combinator Interview," A Really Bad Idea blog, November 24, 2010, http://areallybadidea.com/34320844.
9. Kan, "Y Combinator Interview."

CHAPTER 2: OLDER

1. PG, "A Student's Guide to Startups," October 2006, http://paulgraham.com/mit.html.
2. PG uses "ramen" only as a figurative term and takes pains to say that he does not prescribe an unhealthy diet of instant ramen. Rice and beans were his recommendation for healthy and inexpensive eating. PG, "Ramen Profitable," July 2009, www.paulgraham.com/ramenprofitable.html.
3. PG, "How Not to Die," August 2007, http://paulgraham.com/die.html.
4. PG, "The 18 Mistakes That Kill Startups," October 2006, www.paulgraham.com/startupmistakes.html.
5. Jessica Livingston, *Founders at Work: Stories of Startups' Early Days* (Berkeley, CA: Apress, 2007), 205–22. This is the basis for the remainder of this account of Viaweb, other than specific exceptions noted.
6. Biographical details are drawn from PG's biography on the Viaweb Web site, preserved at http://ycombinator.com/viaweb/com.html.
7. John Markoff, "How a Need for Challenge Seduced Computer Expert," *New York Times*, November 6, 1988, www.nytimes.com/1988/11/06/us/how-a-need-for-challenge-seduced-computer-expert.html; John Markoff, "Computer Intruder Is Put on Probation and Fined $10,000," *New York Times*, May 5, 1990, www.nytimes.com/1990/05/05/us/computer-intruder-is-put-on-probation-and-fined-10000.html. Coverage of the incident did not fail to note that this Robert Morris—Robert Tappan Morris—was the son of Robert Morris, the well-known computer scientist who was then the chief scientist for the National Computer Security Center, a division of the National Security Agency. Many years earlier, Morris fils had had another run-in with university authorities while an undergraduate student at Harvard. In 1988, Morris worked on restoring Harvard's then dormant network connection to Arpanet for no reason other than the intellectual challenge. PG described it this way: "Because he spent all his time on it and neglected his studies, he was kicked out of school for a year." "Undergraduation," March 2005, www.paulgraham.com/college.html.
8. PG, "Student's Guide." When PG and Morris started Viaweb, Morris was so averse to publicity after the worm that he insisted that a pseudonym be used for his biography on the company Web site: he was "John McArtyem," whose real identity was easily seen in the "rtm" in the "rtm@viaweb.com" e-mail address that was displayed. http://ycombinator.com/viaweb/com.html. Many years later, Morris softened a little bit, allowing PG to refer elliptically to the worm incident in the biography provided for Morris on YC's Web site: "In 1988 his discovery of buffer overflow first brought the Internet to the attention of the general public." YC Web site, http://ycombinator.com/people.html.

9. PG, "How Y Combinator Started," YC Web site, March 15, 2012, http://ycombinator.com/start.html.
10. PG, "How to Start a Startup," March 2005, http://paulgraham.com/start.html. In January 2012, PG published a brief reminiscence about Viaweb, PG, "Snapshot: Viaweb, June 1998," http://paulgraham.com/vw.html, and posted a snapshot of the Viaweb site taken in 1998 on the eve of the Yahoo acquisition: http://ycombinator.com/viaweb/.
11. PG later wrote, "Trevor graduated at about the same time the acquisition closed, so in the course of four days, he went from impecunious grad student to millionaire Ph.D." PG, "Snapshot: Viaweb."
12. Livingston, *Founders at Work*, 217.
13. PG, "How Y Combinator Started."
14. PG, "Why Smart People Have Bad Ideas," April 2005, www.paulgraham.com/bronze.html.
15. Perhaps I, a Silicon Valley resident, am inclined to view the South as more isolated from tech centers as a matter of reflex. Richard Florida argues that the South does not receive due credit for attracting venture capital and shows other evidence that it too has become a hospitable host of "Startup Nation." Richard Florida, "The Spread of Start-Up America and the Rise of the High-Tech South," *The Atlantic*, October 2011, www.theatlantic.com/technology/archive/2011/10/the-spread-of-start-up-america-and-the-rise-of-the-high-tech-south/246916/.
16. Chris Dixon, "Selling Pickaxes During a Gold Rush," Chris Dixon blog, February 5, 2011, http://cdixon.org/2011/02/05/selling-pickaxes-during-a-gold-rush/.
17. Robin Wauters, "Salesforce.com Buys Heroku for $212 Million in Cash," TC, December 8, 2010, http://techcrunch.com/2010/12/08/breaking-salesforce-buys-heroku-for-212-million-in-cash/. Heroku had raised only $13 million in capital prior to its sale, so its investors enjoyed outstanding returns in a very short period of time.

CHAPTER 3: GRAD SCHOOL

1. Matt Brezina, "YC: The New Grad School," Matt Brezina blog, April 14, 2011, www.mattbrezina.com/blog/2011/04/yc-the-new-grad-school/.
2. PG, "Student's Guide."
3. PG wanted to set up YC in Berkeley, but he did not have time to find a suitable building there. Being forced to use the space that Blackwell offered in Mountain View turned out well, PG would later say: "We lucked out because Mountain View turned out to be the ideal place to put something like YC." PG, "How Y Combinator Started."
4. PG, "California Year-Round," YC Web site, January 2009, http://ycombinator.com/ycca.html.

5. In 2012, TechStars invests a flat $18,000 for each team; when it began in 2007, the amount was $15,000. Michael Arrington, "TechStars: Summer Camp (and Cash) for Entrepreneurs," TC, January 25, 2007, http://techcrunch.com/2007/01/25/techstars-summer-camp-for-entrepreneurs/.
6. When Paul Buchheit pointed to TechStars' copying the questions, he was writing as an observer of startups and seed funds; this was three years before he became a YC partner. Paul Buchheit, "Did Anyone Else Notice That TechStars and Y Combinator Have the Same Application?" Paul Buchheit blog, March 26, 2007, http://paulbuchheit.blogspot.com/2007/03/anyone-else-notice-that-techstars-and-y.html; discussion on HN, http://news.ycombinator.com/item?id=6505.
7. The list of accelerators comes from Frank Gruber, "Top 15 U.S. Startup Accelerators and Incubators Ranked; TechStars and Y Combinator Top Rankings," Tech Cocktail, May 2, 2011, http://techcocktail.com/top-15-us-startup-accelerators-ranked-2011-05.
8. Comment about TechStars was made by Bart Ciak at a panel organized during Entrepreneur Week in Waterloo, Ontario. At the same panel, Vidyard's Michael Litt, of YC's summer 2011 class, and Garry Tan, a YC partner, also appeared. "California Incubator Not All Sunshine for Local Entrepreneur," *The Record*, November 8, 2011, www.therecord.com/news/business/article/621927.
9. PG, "What I Did This Summer," October 2005, http://paulgraham.com/sfp.html.
10. At the beginning of 2012, one year after the Start Fund made $150,000 investments in YC's winter 2011 batch, TechStars was able to offer a convertible note to every company that was funded. It was a $100,000 note, however, not a $150,000 note, and while the Start Fund's note was uncapped, which let the market set the valuation of the company whenever the debt would be converted into equity, the TechStars' note did have a cap. Each company had to hammer out the terms individually.

CHAPTER 4: MALE

1. PG, "How to Start a Startup."
2. PG, "Ideas for Startups," October 2005, http://paulgraham.com/ideas.html. Seven years later, PG published a new list of ideas: PG, "Frighteningly Ambitious Startup Ideas," March 2012, http://paulgraham.com/ambitious.html.
3. Shira Ovide, "Addressing the Lack of Women Leading Tech Start-Ups," *Wall Street Journal*, August 27, 2010, http://blogs.wsj.com/venturecapital/2010/08/27/addressing-the-lack-of-women-leading-tech-start-ups/.
4. Michael Arrington, "Too Few Women in Tech? Stop Blaming the Men," TC, August 28, 2010, http://techcrunch.com/2010/08/28/women-in-tech-stop-blaming-me/.
5. Jessica Livingston, "What Stops Female Founders?" Founders at Work blog, Janu-

ary 26, 2011, www.foundersatwork.com/1/post/2011/01/what-stops-female-founders.html.

6. A salary survey undertaken by Stanford's computer science department found that among the 140 undergraduates majoring in computer science or electrical engineering in the class of 2011 who responded, salary offers ranged from $64,400 to $100,000 and the median offer was $82,200. Seventy percent were offered stock options and 80 percent were offered signing bonuses. "Stanford Computer Science '10–'11 Salary Survey Results," HN, October 21, 2011, http://news.ycombinator.com/item?id=3141716.
7. Justin Vincent and Jason Roberts, "TechZing 26—Jessica Mah of inDinero," TechZing Tech Podcast, December 10, 2009, www.techzinglive.com/page/146/techzing-26-jessica-mah-of-indinero.
8. Justin Vincent and Jason Roberts, "TechZing 66—Jessica Mah and the Y Combinator Experience," TechZing Tech Podcast, September 9, 2010, www.techzinglive.com/page/409/techzing-66—jessica-mah-the-y-combinator-experience.
9. Jessica Mah, "Culture and Purpose from the Start," Jessica Mah Meets World blog, June 11, 2010, http://jessicamah.com/blog-23-1. The quotation comes from the final paragraph of the long post. A slightly different version of the same paragraph is included earlier in the post.
10. E. B. Boyd, "Where Is the Female Mark Zuckerberg?" *San Francisco*, December 2011 [posted online November 22, 2011], www.modernluxury.com/san-francisco/story/where-the-female-mark-zuckerberg. Earlier in the year, Aileen Lee, a partner at Kleiner Perkins Caufield & Byers, had published on TC a widely read essay, "Why Women Rule the Internet," in which she argued that a new batch of e-commerce companies were springing up that addressed female consumers and that a majority of users of Facebook, Zynga, Groupon, and Twitter were female. Aileen Lee, "Why Women Rule the Internet," TC, March 12, 2011, http://techcrunch.com/2011/03/20/why-women-rule-the-internet/.
11. "Y Combinator's Graham Discusses Start-Up Industry," Bloomberg, videotaped interview with Emily Chang, December 20, 2011, www.bloomberg.com/video/83135286/.

CHAPTER 5: CRAZY BUT NORMAL

1. PG, "What the Bubble Got Right," September 2004, www.paulgraham.com/bubble.html.
2. PG, "Why Startups Condense in America," May 2006, www.paulgraham.com/america.html.
3. In an "Ask Me Anything" forum in the summer of 2011, HT was asked if YC would ever run a program in New York. He wrote, "I wouldn't completely rule out us ever doing something in NY, if we *had* to open another YC branch it would almost

certainly be in NY. The difficulty is in making sure we keep the quality of the experience high, we're still experimenting and trying new things with YC in the Valley to achieve that aim. What makes YC special is the people involved, we wouldn't want to sacrifice that just so we could say we had a NY branch." HT, "I'm a Partner at Y Combinator. Ask Me Anything," AnyAsq, n.d. [summer 2011], http://anyasq.com/29-im-a-partner-at-y-combinator.

4. HT, HN London Meetup, September 29, 2011, http://vimeo.com/30800728. The account that follows is largely based on this source and on "The Year That Made Me: Kulveer Taggar," interview by David Langer, Freed from The Matrix blog, January 23, 2009, http://davidlanger.co.uk/2009/01/23/the-year-that-made-me-kulveer-taggar/.
5. PG, "18 Mistakes."
6. HT, "What I Expected from YC and What I Got," Meal Ticket blog, April 15, 2007, http://mealticket.wordpress.com/2007/04/15/what-i-expected-from-yc-and-what-i-got/.
7. HT, "The Lessons I've Learnt During Y Combinator," Meal Ticket blog, March 11, 2007, http://mealticket.wordpress.com/2007/03/11/the-lessons-ive-learnt-during-y-combinator/.
8. Andrew Warner interview of Jessica Livingston, "How the Author of *Founders at Work* Helps Y Combinator Discover and Mentor Startups—with Jessica Livingston," Mixergy, April 19, 2010, http://mixergy.com/y-combinator-jessica-livingston-interview/.
9. HT, "First Week in 'Frisco,'" Meal Ticket blog, January 15, 2007, http://mealticket.wordpress.com/2007/01/15/first-week-in-frisco/.
10. HT, "The Second Week," Meal Ticket blog, January 22, 2007, http://mealticket.wordpress.com/2007/01/22/the-second-week/.
11. HT, "And Then There Were Three," January 28, 2007, http://mealticket.wordpress.com/2007/01/28/and-then-there-were-three.
12. HT, "Second Week."
13. HT, "What I Expected."
14. HT, "Demo Day," Meal Ticket blog, February 13, 2007, http://mealticket.wordpress.com/2007/02/13/demo-day/.
15. HT, "What I Expected."
16. HT, "What I Expected."
17. Patrick Collison, "Surprises," Patrick Collison blog, October 18, 2009, http://collison.ie/blog/2009/10/surprises.
18. Steven Levy, "Taking the Millions Now," *Newsweek*, April 5, 2008, www.newsweek.com/2008/04/05/taking-the-millions-now.html.
19. "Graduate Entrepreneurs Sell Business for Millions," University of Oxford press release, May 7, 2008, www.ox.ac.uk/media/news_stories/2008/080507b.html.
20. HT, "Leaving Live Current and Vancouver," HT blog, September 5, 2009, http://blog.harjtaggar.com/leaving-live-current-and-vancouver.

21. HT, "Post-Startup School Thoughts," HT blog, October 6, 2009, http://blog.harjtaggar.com/post-startup-school-thoughts.
22. HT, "Auctomatic Is Acquired. Thank You Everyone Who Helped," HT blog, March 27, 2008, http://blog.harjtaggar.com/auctomatic-is-acquired-thank-you-everyone-who.
23. HT answering Quora question: "What Does Harjeet Taggar's Role at Y Combinator Entail, and How Did He Become Partner at 25?" Quora, September 25, 2011, www.quora.com/What-does-Harjeet-Taggars-role-at-Y-Combinator-entail-and-how-did-he-become-partner-at-25.
24. "Y Combinator Announces Two New Partners, Paul Buchheit and Harj Taggar," YC Posterous, November 12, 2010, http://ycombinator.posterous.com/y-combinator-announces-two-new-partners-paul.
25. HT, "I'm a Partner at Y Combinator."
26. "Welcome Sam, Garry, Emmett, and Justin," YC Posterous, June 18, 2011, http://ycombinator.posterous.com/welcome-sam-garry-emmett-and-justin.
27. The official announcement of Aaron Iba's appointment and Garry Tan's promotion to full-time partner was not made until the winter 2012 batch had started. "Welcome Garry and Aaron," YC Posterous, January 23, 2012, http://ycombinator.posterous.com/welcome-garry-and-aaron.
28. The following account is based on an interview that Patrick Collison did with Justin Vincent and Jason Roberts, "TechZing 168—Patrick Collison/Stripe," TechZing Tech Podcast, February 2, 2012, techzinglive.com/page/939/168-tz-interview-patrick-collison-stripe, and Collison's talk at Startup Bootcamp, MIT, September 24, 2011, www.youtube.com/watch?v=M48NAsKE9xY.
29. PG, "Schlep Blindness," January 2012, http://paulgraham.com/schlep.html.
30. "Stripe Said to Get Funding Valuing Online-Payment Startup at $100 Million," Bloomberg, February 9, 2012, www.bloomberg.com/news/2012-02-09/stripe-said-to-get-funding-valuing-online-payment-startup-at-100-million.html.
31. Kulveer Taggar, "Moving Back to SF and Doing Y Combinator Again," Kulveer Taggar blog, June 7, 2011, http://kulveer.co.uk/2011/06/07/moving-back-to-sf-and-doing-y-combinator-again/.

CHAPTER 6: UNSEXY

1. PG, "Why Startup Hubs Work," October 2011, http://paulgraham.com/hubs.html.

CHAPTER 7: GENIUS

1. PG conceded that it was possible to launch too fast, ruining the startup's reputation. But when PG wrote "The 18 Mistakes That Kill Startups" in October 2006, he said

that "launching too slowly has probably killed a hundred times more startups than launching too fast." He advised founders to trust that early adopters will be fairly tolerant of incomplete products. They "don't expect a newly launched product to do everything; it just has to do *something*."

2. Eric Ries, "Building the Minimum Viable Product," Entrepreneurial Thought Leader Lecture Series, Entrepreneurship Corner, Stanford University, September 30, 2009, http://ecorner.stanford.edu/authorMaterialInfo.html?mid=2295.
3. Steve Blank, "Perfection by Subtraction—The Minimum Feature Set," Steve Blank blog, March 4, 2010, http://steveblank.com/2010/03/04/perfection-by-subtraction-the-minimum-feature-set/.
4. Reid Hoffman, remarks at the Churchill Club's "Startup Success 2006," August 17, 2006, http://video.google.com/videoplay?docid=2401538119328376288. Hoffman actually said, "If you're not embarrassed in consumer Internet by the first version of the product you've launched, you've launched too late," but in the retelling by others, the qualifier "in consumer Internet" was dropped.
5. "Clustrix Emerges from Stealth Mode with Industry's First Clustered Database System for Internet-Scale Applications," Clustrix press release, May 4, 2010, www.clustrix.com/company/news-events/press-releases/bid/82423/Clustrix-Emerges-From-Stealth-Mode-With-Industry-s-First-Clustered-Database-System-for-Internet-Scale-Applications.
6. PG said he first heard the phrase from Joe Kraus, who told PG that he believed it had originated with either William Hewlett or David Packard. I have not been able to confirm the attribution.
7. "What Is Rap Genius?" Rap Genius Web site, http://rapgenius.com/static/about.

CHAPTER 8: ANGELS

1. PG, "The Hacker's Guide to Investors," April 2007, www.paulgraham.com/guidetoinvestors.html.
2. Fred Wilson, "Recycling Capital," AVC blog, April 17, 2011, www.avc.com/a_vc/2011/04/reinvesting-capital.html.
3. Dan Primack, "Exclusive: SV Angel's Investment Portfolio," *Fortune*, November 22, 2011, http://finance.fortune.cnn.com/2011/11/22/exclusive-sv-angels-investment-portfolio/. For a portrait of Conway during the dot-com boom of the late 1990s, see Gary Rivlin, *The Godfather of Silicon Valley: Ron Conway and the Fall of the Dot-coms* (New York: AtRandom, 2001); for a contemporary portrait, see Miguel Helft, "Ron Conway Is a Silicon Valley Startup's Best Friend," *Fortune*, February 10, 2012, http://tech.fortune.cnn.com/2012/02/10/ron-conway-sv-angel/.
4. Dave McClure interview, "9th Founder Showcase—Alexia Tsotsis of TechCrunch Interviews Dave McClure of 500 Startups," http://vimeo.com/35399949; Anthony Ha, "Dave McClure Isn't Worried About the 'Series A Crunch,'" TC,

January 21, 2012, http://techcrunch.com/2012/01/21/dave-mcclure-series-a-crunch/.

CHAPTER 9: ALWAYS BE CLOSING

1. In Chris Tam's class at Harvard Business School, one other member went directly into YC's summer batch: Streak's Aleem Mawani.
2. The line was delivered with memorable malevolence by Alec Baldwin in the 1992 film adaptation of *Glengarry Glen Ross*.
3. Clay Shirky, "The Case Against Micropayments," OpenP2P, December 19, 2000, http://openp2p.com/pub/a/p2p/2000/12/19/micropayments.html. Shirky took issue in particular with the advocacy of Jakob Nielsen's "The Case for Micropayments," Alertbox blog, January 25, 1998, www.useit.com/alertbox/980125.html.
4. Calvin Young, "Minno Makes a Splash," BuySimple blog, March 30, 2011, http://blog.buysimple.com/2011/03/30/minno-makes-a-splash-2/. See also Joe Mullin, "Ex-Googlers Launch 'NYT for a Nickel' as Publicity Stunt; NYT Not Amused," PaidContent, March 28, 2011, http://paidcontent.org/article/419-ex-googlers-launch-nyt-for-a-nickel-as-publicity-stunt-nyt-not-amused/. The *Times* said it did not want the nominal $33.85 that had been collected, but the Minno founders sent the newspaper a check for that amount anyway.

CHAPTER 11: WHAT'S UP?

1. Jason Shen, "The Rejection Therapy Challenge: Week 1," Art of Ass-Kicking blog, October 20, 2011, www.jasonshen.com/2010/the-rejection-therapy-challenge-week-1/. Shen's gauntlet of rejection was covered in the *San Francisco Chronicle*: Meredith May, "Experimenting with Rejection Builds Self-Confidence," *San Francisco Chronicle*, November 27, 2010, http://articles.sfgate.com/2010-11-27/news/24948328_1_rejection-stings-elayne-savage-social-media. For the simple rules of rejection therapy, see http://rejectiontherapy.com/rules/.
2. When Holden and Lynch decide to close Splitterbug in mid-July, mere days after announcing its beta release, one commentator on HN asked why PG had funded the idea, implicitly suggesting that its limitations would have been as obvious at the beginning of the session as six weeks later. PG posted an answer to the query: "This wasn't even the idea they applied with." "Splitterbug (YC S11) Shutting Down," HN, July 13, 2011, http://news.ycombinator.com/item?id=2759880.
3. The two founders, Noah Ready-Campbell and Calvin Young, subsequently came up with a new idea to work on: a Web site for selling secondhand clothing, Like Twice.com.

CHAPTER 12: HACKATHON

1. Jason Kincaid, "YC-Funded Snapjoy Will Organize Your Photos for You (And Make Sure You Don't Lose Them)," TC, August 8, 2011, http://techcrunch.com/2011/08/08/yc-funded-snapjoy-will-organize-your-photos-for-you-and-make-sure-you-dont-lose-them/.
2. Where YC did not publicly disclose these, TechStars, which had arranged similar arrangements for its startups, used the list on its Web site as part of its marketing; it said in February 2012 that the total value of the noncash "perks" for each startup was $154,500, separate from the $100,000 convertible note offered to every company funded. www.techstars.com/program/perks/.
3. PG, "How to Start a Startup."
4. PG, "What Business Can Learn from Open Source," August 2005, http://paulgraham.com/opensource.html.
5. PG, "What Business Can Learn."
6. PG urged founders to avoid coworking space such as RocketSpace. One evening, he said, "It's got to be your own space. You can't just be desks 23 through 26. It's really difficult to imagine Larry and Sergey working in some coworking space at Google."

CHAPTER 13: NEW IDEAS

1. "The Thiel Fellowship: 20 Under 20," Thiel Foundation press release, September 29, 2010, www.thielfellowship.org/wp-content/uploads/2011/10/The-Thiel-Fellowship-20-Under-20.pdf. Thiel earned a bachelor's degree in 1989 and a law degree in 1992, both at Stanford. The year after the Thiel Fellowship was announced, he contacted Stanford to teach a course in Stanford's computer science department. It was approved for the spring 2012 quarter after due notice was taken by the Stanford faculty of "what he's said in the past about the value of a university education," said a Stanford faculty member. "Peter Thiel to Teach Stanford Class on Startups," Tech Chronicles blog, *San Francisco Chronicle*, March 12, 2012, http://blog.sfgate.com/techchron/2012/03/12/peter-thiel-to-teach-stanford-class-on-startups/.
2. Justin Kan, "Drop Out. Or Don't," A Really Bad Idea blog, February 27, 2011, http://areallybadidea.com/drop-out-or-dont.
3. Justin Kan, "Selling Kiko," A Really Bad Idea blog, February 21, 2011, http://areallybadidea.com/selling-kiko.
4. Justin Kan, "Why Starting Justin.tv Was a Really Bad Idea, but I'm Glad We Did It Anyway," TC, February 12, 2011, http://techcrunch.com/2011/02/12/starting-justin-tv/. This account of Justin.tv's history and its double pivot is also based on the author's interview with Kan and Shear, December 29, 2011.
5. John Gaudiosi, "Pro Gamer Tyler 'Ninja' Blevins Discusses Meteoric Rise of Major League Gaming," *Forbes*, December 6, 2011, www.forbes.com/sites/johngaudiosi/2011/12/06/pro-gamer-tyler-ninja-blevins-discusses-meteoric-rise-of-major-

league-gaming/. Blevins said another player, "Destiny," had told him he was making $450 a day. These figures were mentioned in December 2011 and undoubtedly were larger than they would have been in June 2011 when Twitch.tv launched.

6. Jason Kincaid, "Socialcam 2.0 Lands on the iPhone," TC, April 20, 2011, http://techcrunch.com/2011/04/20/socialcam-2-0-lands-on-the-iphone/.
7. "AARRR!" was used in a talk McClure gave in 2007, summarized in his blog post "Product Marketing for Pirates: AARRR! (aka Startup Metrics for Internet Marketing & Product Management)," Master of 500 Hats blog, June 20, 2007, http://500hats.typepad.com/500blogs/2007/06/internet-market.html. AARRR was an acronym for Acquisition, Activation, Retention, Referral, and Revenue.

CHAPTER 14: RISK

1. "Where Are They Now: Ralston Shepherds Yahoo E-mail from Free to Paid," MarketWatch, September 20, 2002, www.marketwatch.com/story/the-man-in-charge-of-yahoo-e-mail-shares-his-vision.
2. Apple did not rebrand Lala's service; it shut the service down the next year. Peter Kafka, "Apple Pulls the Plug on Lala, Replaces It with . . . Nada," AllThingsD, June 1, 2010, http://allthingsd.com/20100601/apple-pulls-the-plug-on-lala-replaces-it-with-nothing/.
3. PG, "Imagine K12," YC Web site, March 17, 2011, http://ycombinator.com/imagineK12.html.
4. Ralston was formally appointed a YC partner early the next year. "Welcome Geoff," YC Posterous, January 27, 2012, http://ycombinator.posterous.com/welcome-geoff.
5. The CueCat received the dubious honor of being included in *PC World*'s list of "The 25 Worst Tech Products of All Time," *PC World*, May 26, 2006, www.pcworld.com/article/125772-8/the_25_worst_tech_products_of_all_time.html.
6. PG, "Smart People." PG had first considered instead printing the following on the backs of the shirts: "If you can read this, I should be working."
7. PG, "The Future of Web Startups," October 2007, www.paulgraham.com/webstartups.html.
8. PG, "How to Make Wealth," May 2004, http://paulgraham.com/wealth.html.

CHAPTER 15: MARRIED

1. Andrew Warner interview of PG, "How Y Combinator Helped 172 Startups Take Off—With Paul Graham," Mixergy, February 10, 2010, http://mixergy.com/y-combinator-paul-graham/.
2. PG, "18 Mistakes."
3. HT, HN London Meetup.
4. PG said, "Cofounders are for a startup what location is for real estate. You can

change anything about a house except where it is. In a startup you can change your idea easily, but changing your cofounders is hard. And the success of a startup is almost always a function of its founders." PG, "Startups in 13 Sentences," February 2009, www.paulgraham.com/13sentences.html.

5. PG, "What We Look For in Founders," October 2010, www.paulgraham.com/founders.html.
6. PG, "What Startups Are Really Like," October 2009, http://paulgraham.com/really.html. This was based on a talk presented at the 2009 Startup School. PG surveyed YC alumni, asking what surprised them about starting a startup.
7. PG, "Student's Guide."
8. PG, "What We Look For."
9. Jason Shen, "How to Find Awesome Startup Roommates," Art of Ass-Kicking blog, February 22, 2011, www.jasonshen.com/2011/how-to-find-awesome-startup-roommates/. Shen's post came two years after the search that had brought in Randy Pang and encompassed two other successful searches as well.
10. Kalvin Wang, "How Borderline Douchebaggery Helps You Land a Great Roommate," Tech & Do-Goodery blog, February 21, 2011, http://kalv.in/how-borderline-douchebaggery-helps-you-land-a-great-roommate/. The next day, Wang noted that the actual Jason Shen should be distinguished from the online persona, and the real one is "awesome." Kalvin Wang, "The Difference Between Jason Shen and Jason Shen.com," Tech & Do-Goodery blog, February 22, 2011, http://kalv.in/the-difference-between-jason-shen-and-jasonshen-com/.
11. http://apps.facebook.com/graffitiwall.
12. For samples of some extraordinary Minecraft art, such as a rendering of the Taj Mahal, see David Thomas, "How the Creator of Minecraft Developed a Monster Hit," *Wired*, December 2011, www.wired.com/magazine/2011/11/st_alphageek_minecraft/.
13. HT, "I'm a Partner at Y Combinator." Asked what is "the hardest part" of his position at YC, HT said reading the applications, which was "draining in a way I've never experienced before." Giving each application his full attention, while reading hundreds, "just fries my brain." Far easier, he said, was conducting the interviews, where he could talk to the applicants.
14. Excelerate Labs, which began in 2010, invested $25,000 in each of the ten companies in its 2011 batch, in exchange for a 6 percent share of equity. In 2011, New World Ventures offered $50,000 in the form of convertible debt to each startup in the program. www.exceleratelabs.com/details/.

CHAPTER 16: FEARSOME

1. At the beginning of the summer, this startup used the domain name of TheBench.com; it switched to ScienceExchange.com in August. Iorns's Kiwi accent caused

"TheBench" to be misheard in at least one instance. She reported during office hours that one person who had heard her say "TheBench" thought she had said, "TheBitch." Iorns followed her recounting with "Oh, my goodness!"

CHAPTER 17: PAY ATTENTION

1. When Kulveer Taggar went through YC in 2007, he had seven minutes to make the pitch on Demo Day, but he likes the shorter format: "The 2.5 minute pitches are also laser focused. I now struggle to think what we talked about for 7 minutes back in 2007." "Doing Y Combinator a Second Time," Kulveer Taggar blog, January 28, 2012, http://kulveer.co.uk/2012/01/28/doing-y-combinator-a-second-time/.
2. Steiner is the only YC founder who wrote about YC for a major publication and then became a YC founder himself. See Christopher Steiner, "The Disruptor in the Valley," *Forbes*, November 8, 2010, www.forbes.com/forbes/2010/1108/best-small-companies-10-y-combinator-paul-graham-disruptor.html.

CHAPTER 18: GROWTH

1. Jason Kincaid, "Interview Street Streamlines the Search for Great Programmers," TC, August 6, 2011, http://techcrunch.com/2011/08/06/yc-funded-interview-street-streamlines-the-search-for-great-programmers/; Alexia Tsotsis, "YC-Backed Leaky Is Hipmunk for Car Insurance," TC, August 8, 2011, http://techcrunch.com/2011/08/08/yc-backed-leaky-is-hipmunk-for-car-insurance/; Kincaid, "YC-Funded Snapjoy Will Organize Your Photos"; Jason Kincaid, "YC-Funded Stypi Is Etherpad Reborn," TC, August 9, 2011, http://techcrunch.com/2011/08/09/yc-funded-stypi-is-etherpad-reborn/; Jason Kincaid, "YC-Funded Envolve Launches an API for Real-Time Chat," TC, August 10, 2011, http://techcrunch.com/2011/08/10/yc-funded-envolve-launches-an-api-for-real-time-chat/; Jason Kincaid, "YC-Funded MobileWorks Aims to Be a Hands-Off Mechanical Turk," TC, August 12, 2011, http://techcrunch.com/2011/08/12/yc-funded-mobileworks-aims-to-be-a-hands-off-mechanical-turk/; Sarah Perez, "YC-Funded Picplum: Beautiful Prints, Automatically Mailed for You," August 12, 2011, http://techcrunch.com/2011/08/12/yc-funded-picplum-beautiful-prints-automatically-mailed-for-you/.
2. "Show HN: Codecademy.com, the Easiest Way to Learn to Code," HN, August 18, 2011, http://news.ycombinator.com/item?id=2901156.
3. "Learn to Code: Codecademy," LearnProgramming, Reddit, August 18, 2011, www.reddit.com/r/learnprogramming/comments/jniah/learn_to_code_codecademy_xpost_from_rprogramming/.
4. Jason Kincaid, "Codecademy: A Slick, Fun Way to Teach Yourself How to Program," TC, August 18, 2011, http://techcrunch.com/2011/08/18/codecademy-a-slick-fun-way-to-teach-yourself-how-to-program/.

5. PG, "Student's Guide."
6. Paul Buchheit, "The Most Important Thing to Understand About New Products and Startups," Paul Buchheit blog, February 17, 2008, http://paulbuchheit.blogspot.com/2008/02/most-import-thing-to-understand-about.html.
7. Livingston, *Founders at Work*; Paul Buchheit, "Serendipity Finds You," Paul Buchheit blog, October 24, 2010, http://paulbuchheit.blogspot.com/2010/10/serendipity-finds-you.html.
8. Buchheit was one of four cofounders of FriendFeed, which was acquired by Facebook in 2009 for an undisclosed amount. In January 2012, when Facebook released its S-1 for its IPO, the Silicon Alley Insider matched up the timing of the FriendFeed acquisition with Facebook's issuing more than eleven million shares of Class B common stock related to an acquisition. Using an internal valuation used by Facebook the previous month, the tech business blog estimated the value of shares issued to FriendFeed's owners to be $328 million. Alyson Shontell, "Now We Know How Many Millions of Dollars These Startups Made Selling to Facebook," SAI Business Insider, February 2, 2012, www.businessinsider.com/facebook-acquisition-shares-stock-startups-2012-2.
9. Paul Buchheit, "Angel Investing: My First Three Years," Paul Buchheit blog, January 3, 2011, http://paulbuchheit.blogspot.com/2011/01/angel-investing-my-first-three-years.html.
10. Buchheit, "Angel Investing."
11. In 1982, the SEC set forth specific qualifying amounts—a net worth exceeding $1 million, or annual income greater than $200,000 in the previous two years, with some fine print. It also offered exemptions to others, such as directors or executives in a company, or to "sophisticated investors" who have access to the type of information normally provided in a prospectus and who agree not to resell the securities to the public. But the agency did not offer the same bright-line rules about who qualifies as a "sophisticated investor" as it did for "accredited investors," so most companies preferred to sell privately placed securities only to the latter category of investors.
12. The JOBS Act included a crowdfunding exemption from the standard restrictions governing equity sales to accredited investors. This allowed "emerging growth companies" to raise up to $1 million in a twelve-month period in equity sales to investors who were not accredited investors. These investors could purchase the greater of $2,000 or 5 percent of his or her annual income or net worth, if either was $100,000 or less; or no more than 10 percent if over $100,000. www.gpo.gov/fdsys/pkg/BILLS-112hr3606enr/pdf/BILLS-112hr3606enr.pdf.

CHAPTER 19: FIND A DROPBOX

1. It was a matter of public record that Ashton Kutcher had invested in more tech companies—including Foursquare, Skype, Path, Flipboard, and YC-backed Airbnb—

than any other actor, earning him a *New York Times* profile as an investor: Jenna Wortham, "An Actor Who Knows Start-Ups," *New York Times*, May 25, 2011, www.nytimes.com/2011/05/26/technology/26ashton.html. Two weeks after YC's summer 2011 Demo Day, in an interview at TC Disrupt in September 2011, Kutcher said that he had invested in forty startups to date. Alexia Tsotsis, "Ashton Kutcher: Good Investors Are on a Witch Hunt," TC Disrupt, September 13, 2011, http://techcrunch.com/2011/09/13/ashton-kutcher-good-investors-are-on-a-witch-hunt/.

2. Kulveer Taggar, "Doing Y Combinator a Second Time."
3. Investors did not get to see what the founders saw at an informal YC party a few days earlier, when Shen performed ten handstand push-ups in a row. Readers who wonder if this is difficult should give a handstand push-up a try. Shen wrote up a long, reflective—and at one point, gory—account of his experiences in competitive gymnastics, describing a devastating injury suffered during a meet in his junior year at Stanford, and his return to competition the following year. "How I Blew Out My Knee and Came Back to Win a National Championship," Art of Ass-Kicking blog, January 19, 2011, www.jasonshen.com/2011/blew-out-knee-win-national-championship/; part two, January 30, 2011, www.jasonshen.com/2011/part-2-blew-out-knee-win-national-championship/; part three, February 4, 2011, www.jasonshen.com/2011/part-3-blew-out-knee-win-national-championship/.
4. Andreessen, "Software Is Eating the World."
5. PG, "Why to Move to a Startup Hub," October 2007, www.paulgraham.com/startuphubs.html.

CHAPTER 20: DON'T QUIT

1. "Ask PG: Can You Please Provide Statistics of YC Funded Cos [Companies]?" HN, April 30, 2008, http://news.ycombinator.com/item?id=177606.
2. Paul Stamatiou, "Notifo (YC W2010) Gets a Co-Founder . . . Me," Paul Stamatiou blog, June 26, 2010, http://paulstamatiou.com/notifo-yc-w2010-gets-a-co-founder-me.
3. Chad Etzel, "Startups Are Hard," JazzyChad blog, May 2, 2011, http://blog.jazzychad.net/2011/05/02/startups-are-hard.html.
4. "Notifo Will Be Shutting Down," Notifo blog, September 8, 2011, http://blog.notifo.com/notifo.
5. PG, "Y Combinator Numbers," June 2011, http://ycombinator.com/nums.html.
6. James Middleton, "Founding Father," Telecoms.com, November 30, 2011, www.telecoms.com/37300/founding-father/.
7. Kim-Mai Cutler, "Zynga No Longer Has the Biggest Game on Facebook by Daily Users. OMGPOP Does," TC, March 16, 2012, http://techcrunch.com/2012/03/16/zynga-omgpop/; Brian Chen and Jenna Wortham, "A Game Explodes and Changes Life Overnight at a Struggling Start-Up," *New York Times*, March 25, 2012, www

.nytimes.com/2012/03/26/technology/draw-something-changes-the-game-quickly-for-omgpop.html; Kim-Mai Cutler, "The Inside Story of the OMGPOP-Zynga Deal from the CEO, Investors and More!" TC, March 21, 2012, http://techcrunch.com/2012/03/21/zynga-omgpop-porter-sabet-david-ko/.

CHAPTER 21: SOFTWARE IS EATING THE WORLD

1. The Code Year Web site displays quotes from Fred Wilson, of Union Square Ventures (and one of Codecademy's investors), who says, "A young man asked me for advice for 'those who aren't technical.' I said he should try to get technical," and from Douglas Rushkoff, the author of *Program or Be Programmed*, who says, "If we don't learn to program, we risk being programmed ourselves." http://codeyear.com/.
2. Alexia Tsotsis, "Still Looking for a New Year's Resolution? How about Learning How to Code . . . ," TC, January 1, 2012, http://techcrunch.com/2012/01/01/new-years-resolution-programming/.
3. Jason Kincaid, "Codecademy's Code Year Attracts 100,000 Aspiring Programmers in 48 Hours," TC, January 3, 2012, http://techcrunch.com/2012/01/03/codecademys-codeyear-attracts-100000-aspiring-programmers-in-48-hours/.
4. Carl Franzen, "Mayor Bloomberg Will Learn How to Code in 2012," Talking Points Memo, January 6, 2012, http://idealab.talkingpointsmemo.com/2012/01/mayor-bloomberg-will-learn-how-to-write-code-in-2012.php.
5. Aneesh Chopra, the first United States chief technology officer, announced that Codecademy would develop an abbreviated version of the Code Year program for the president's "Summer Jobs +" program for low-income young people. "New Summer Jobs + Commitments Plan to Introduce Low-Income Youth to Technology-Related Skills," White House blog, January 17, 2012, www.whitehouse.gov/blog/2012/01/17/new-summer-jobs-commitments-plan-introduce-low-income-youth-technology-related-skill; "Announcing Meetups and Our Partnership with the White House," Codecademy blog, January 17, 2012, http://www.codecademy.com/blog/5-announcing-meetups-and-our-partnership-with-the-white-house.
6. Among the startups in the winter 2012 batch, at least one company, 99dresses, was in the fashion business. Founder Nikki Durkin, twenty, is Australian and her mention of being accepted into YC on Facebook was picked up by the *Sydney Morning Herald*, which described Australian entrepreneurs seeking to go to Silicon Valley as "like religious fanatics trying to get to their mecca." Asher Moses, "Aussie Nikki Joins Silicon Valley Millionaire Factory," November 18, 2011, www.smh.com.au/technology/technology-news/aussie-nikki-joins-silicon-valley-millionaire-factory-20111118-1nlud.html. At the first dinner, PG was in the middle of his standard spiel for technical founders when he remembered 99dresses, so different from the many companies whose founders sell developer tools to fellow developers. He looked over in the direc-

tion of 99dresses and offered a joke to the group: "Conventionally, you're allowed to require that one another use your products. You don't all have to use 99dresses."

7. PG, "How to Be Silicon Valley," May 2006, http://paulgraham.com/siliconvalley.html.
8. PG, "How to Be Silicon Valley."
9. PG, "Can You Buy a Silicon Valley? Maybe," February 2009, http://paulgraham.com/maybe.html. Three months later, PG wrote another essay and clarified that this suggestion had not been intended as a fully fleshed-out, practical proposal, but more of a thought exercise, exploring what would be required, at a minimum, to create a startup hub in a place where one did not exist. PG, "A Local Revolution?," April 2009, http://paulgraham.com/revolution.html.
10. PG, "The Word 'Hacker.'"
11. PG, "Why Startup Hubs Work."
12. Two lists of international startup accelerators are found at Startup Weekend, http://startupweekend.org/incubators/, and at Robert Shedd blog, http://blog.shedd.us/321987608/. For Finland and northern Europe in general, see Startup Sauna, described in "European Startup Accelerators Are Gradually Revealing Their Performance Figures," TC, January 27, 2012, http://techcrunch.com/2012/01/27/european-startup-accelerators-are-gradually-revealing-their-performance-figures/. For Dubai, see SeedStartup, described in Rip Empson, "Founder of Dubai's First Startup Accelerator Looks to Educate, Inspire Global Entrepreneurs," TC, April 4, 2012, http://techcrunch.com/2012/04/04/founder-of-dubais-first-startup-accelerator-looks-to-educate-inspire-global-entrepreneurs/.

INDEX